Mode-Inspired Wideband and Multiband Planar Antennas

For a complete listing of titles in the
Artech House Antennas and Propagation Library,
turn to the back of this book.

Mode-Inspired Wideband and Multiband Planar Antennas

Neng-Wu Liu

Xiao Zhang

Lei Zhu

ARTECH HOUSE

BOSTON | LONDON

artechhouse.com

Library of Congress Cataloging-in-Publication Data
A catalog record for this book is available from the U.S. Library of Congress.

British Library Cataloguing in Publication Data
A catalogue record for this book is available from the British Library.

Cover design by Andy Meaden

ISBN 13: 978-1-68569-035-9

685 Canton Street
Norwood, MA 02062

10 9 8 7 6 5 4 3 2 1

Contents

Preface

Over the past three decades, planar antennas—such as microstrip patch antennas, slot antennas, and printed dipole antennas—have gained popularity due to their inherent advantages, including low profile, compact size, and ease of fabrication. As a result, they have been widely used in applications like cellular communication, satellite communication, and radar systems. However, as modern communication systems evolve to support multifunctionality, multiuser connectivity, and enhanced performance under constrained conditions, traditional planar antennas face limitations because they typically operate using only a single fundamental mode. To address this challenge, the concept of multimode resonance (MMR), originally developed for filters, has been adapted for antennas by research teams worldwide. This approach enables more efficient use and control of both fundamental and higher-order modes, improving impedance and radiation characteristics, and significantly expanding the performance potential of planar antennas.

Compared to other methods, the MMR approach in planar antennas offers four major advantages. First, it enables the use of multiple resonant modes within a single radiator, reducing the number of required antenna elements and lowering the overall antenna thickness. Second, the MMR methodology introduces a novel analytical framework that clarifies the underlying physical principles, thereby improving design efficiency. Third, this approach transforms traditionally unusable higher-order modes into beneficial modes with enhanced performance. Fourth, MMR techniques overcome the performance limitations of conventional planar antennas, often yielding unexpected improvements in radiation characteristics.

Historically, the concept of MMR originated in 1951 with Professor Wei-Gan Lin, who applied it to compact, narrowband waveguide bandpass filters using degenerate modes. Later, in 2005, Professor Lei Zhu extended this theory to single resonators using both fundamental and higher-order modes to develop planar filters with ultrawideband performance. While many planar antennas also exhibit multimode behavior, effectively managing and reusing these modes is more challenging than in filters due to the complex radiation properties introduced by higher-order modes, such as high sidelobes, distorted patterns, and elevated cross polarization.

As a result, MMR-based antenna design has emerged as a key research area in academia and industry, as reflected by a growing body of literature and numerous startups. Our team, Professor Neng-Wu Liu from Xidian University, Professor Xiao Zhang from Shenzhen University, and Professor Lei Zhu from the University of Macau, began research into MMR antennas in 2014, focusing on microstrip patch antennas with advanced performance. Since then, we have published over 100 peer-reviewed journal papers. However, much of this work and related research remains dispersed across various publications. This book aims to consolidate these findings and provide a comprehensive, practical guide for the design of high-performance planar antennas for wireless communication systems.

The structure of this book is as follows:

Chapter 1 introduces fundamental concepts of planar antennas, including input impedance, isolation, gain, and radiation patterns.

Chapter 2 presents the theoretical basis of MMR in planar antennas, covering mode suppression, excitation, frequency tuning, and radiation pattern transformation to improve bandwidth characteristics.

Chapter 3 details MMR techniques for enhancing impedance bandwidth using methods like stepped impedance resonators, shorting pins, slots, and specialized feeding networks.

Chapter 4 explores MMR strategies for improving isolation bandwidth, including the use of even- and odd-mode resonance and mode combinations.

Chapter 5 discusses gain-bandwidth enhancement through techniques for both linear and circular polarization.

Chapter 6 covers pattern-bandwidth enhancement methods, such as widebeam, low sidelobe, and low cross-polarization techniques, as well as beam scanning.

We hope this book serves as a valuable resource for students, researchers, and professionals working on MMR antennas, circuits, phase shifters, and related microwave components.

Finally, we wish to express our sincere gratitude to all authors cited in the references for their foundational contributions. Special thanks goes to Professor Wen-Jun Lu (Nanjing University of Posts and Telecommunications) for his invaluable feedback, and to our PhD and graduate students—Jun-Bing Duan, Rui Du, Yong-Jian Wu, Shu-Wei Yu, Tian-Long Yang, Chang Fang, and

Pin-Cai He—for their technical input. We are also grateful to the Artech House editorial team, including Julia Sayger, Gavin Jamieson, Keri Dickens, Leona Crawford, and Merlin Fox, for their enthusiastic support and guidance. This book would not have been possible without their encouragement and professional assistance.

1

Introduction

1.1 Introduction of Planar Antennas

In modern wireless communications, the planar antenna, which can be made very thin, is in high demand due to the advantages of conformal design options, low-manufacturing cost, and easy fabrication. These factors have led to the design and development of many planar antennas [1]. Commonly used planar antennas include slot antennas [2, 3], printed dipole antennas [4, 5], and microstrip patch antennas (MPAs) [6, 7]. Thin dielectric substrates or air substrates can be used to construct these antennas, with the radiating metals printed on the top or bottom layer of the dielectric substrate.

The radiation patterns of slot antennas and printed dipoles are influenced by the environment, due to insufficient ground plane underneath these radiators. However, MPAs have kept the inherent large ground plane underneath the radiating patch, maintaining relatively stable input impedance and far-zone-radiation pattern performance. Most important, the mode resources of MPAs are richer than other types of planar antennas due to their 2D degrees. As a result, MPAs are widely used in satellite communications, radar detection, remote sensing ranging, missile telemetry, electronic countermeasures, weapon fuses, medical microwave radiometers, and millimeter wave communications. Figure 1.1 shows several well-known application scenarios for the planar antennas.

1.2 Multimode Resonance-Inspired Wideband and Multiband Planar Antennas

Mode analysis is derived and inspired from microwave filters [8]. After considering radiation performance, a number of resonant frequencies are also studied and generated for the planar antennas; the relevant electric-field outlines for these planar antennas are plotted in Figures 1.2, 1.3, and 1.4. For the slot antenna, its first-order, second-order, and third-order modes are plotted in Figure 1.2, which demonstrates that the wavelength of its first-order mode is about twice and three-times longer than that of its second-order and third-

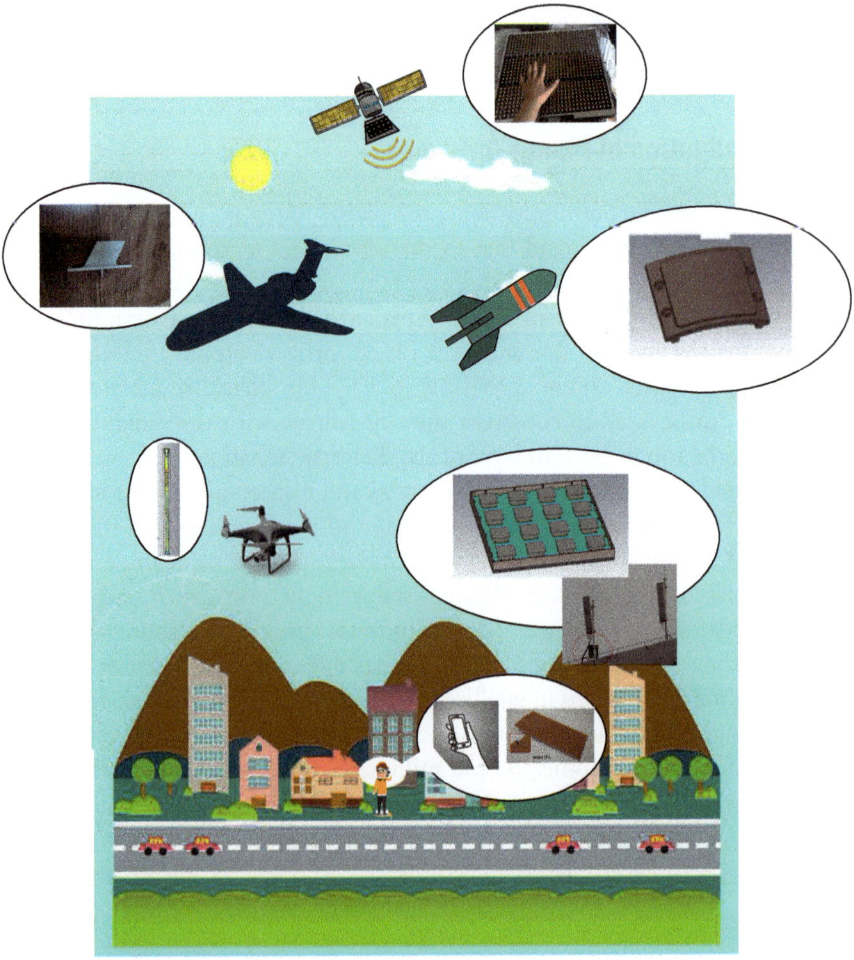

Figure 1.1 Application scenarios of the planar antennas.

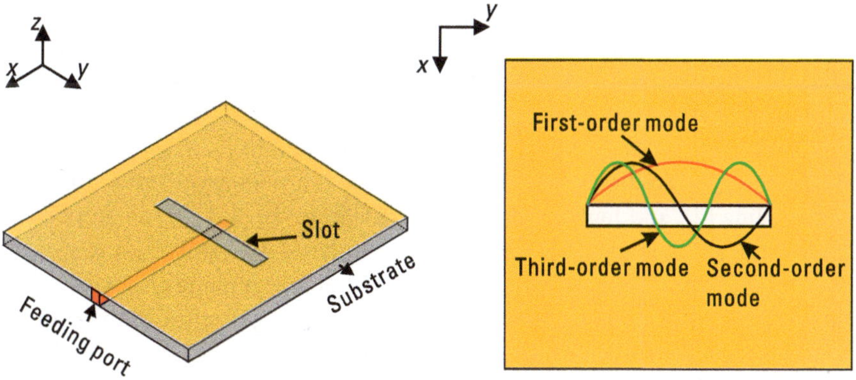

Figure 1.2 The configuration and electric-field-distribution outline of the slot antenna under multimode.

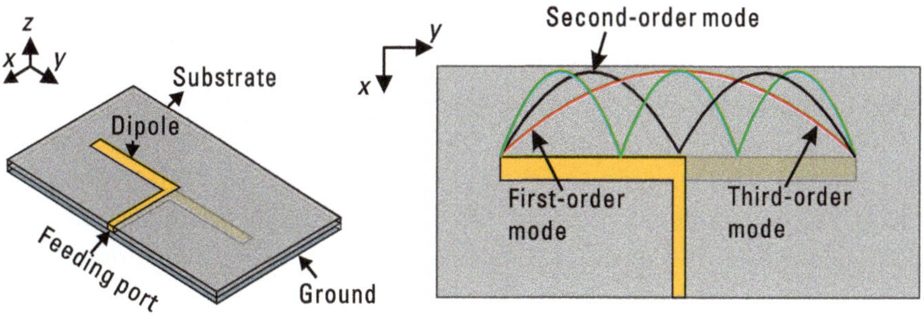

Figure 1.3 The configuration and electric-field-distribution outline of the printed dipole antenna under multimode.

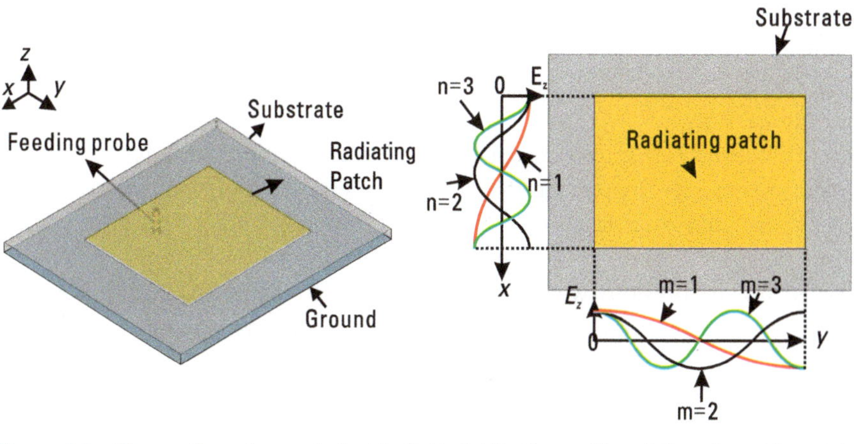

Figure 1.4 The configuration and electric-field-distribution outline of the microstrip patch antenna under multimode resonance.

order modes, respectively. The resonant frequencies of its first-order, second-order, and third-order modes are kept at about f_1, $2f_1$, $3f_1$, independently. The feeding port should be located far away from the electric-field nulls of these radiative modes to excite them effectively. To realize the wide bandwidth or multibandwidth, the previous works of slot antennas in [2, 3] were conducted with improved configuration. By reallocating the microstrip mode from the feeding line [2], and higher third-order mode from the slot [3], the impedance bandwidth of the antennas was significantly widened with two minimum poles.

The electric-field outlines of the printed dipole antenna are presented in Figure 1.3. This figure shows that the antenna also generates the first-order, second-order, and third-order modes during the lower and upper bands [9], and their resonant frequencies are maintained at around f_1, $2f_1$, $3f_1$, respectively. For sufficient excitation of the desired resonant mode(s), a single feed/port should be set at or near the antinode of mode functions. The detail feeding scheme about the mode function's antinode could reference Chapter 1 in [10]. To realize the wide bandwidth or multibandwidth, the previous work of the dipole antenna in [5] was conducted with the improved configuration, which illustrates that circularly polarized (CP) performance is realized for the antenna by simultaneously exciting its orthogonal modes.

The electric-field outlines of MPAs are presented in Figure 1.4. Different from previous slot and dipoles, the electric fields of the MPA are varied along the x-direction and the y-direction. We need to define its radiative modes with two numbers (m and n) to represent these two directions. The number along the z-direction is kept at zero and can be ignored due to small thickness; the feeding port should be located far away from the electric-field nulls of these radiative modes to excite them effectively. The resulting mode resources of the MPAs are richer than those of slot antennas and planar dipole antennas. To realize the wide bandwidth or multibandwidth, the previous work of the MPA in [7] was conducted with the improved configuration, which shows that TM_{10} and TM_{12} modes were successfully excited and reallocated in proximity to each other for wide impedance bandwidth and monopole-like radiation pattern simultaneously.

To efficiently improve the impedance and radiation characteristics of planar antennas, researchers have carried out a series of studies on relative bandwidth around planar antennas with multiple resonant modes, including impedance bandwidth, isolation bandwidth, gain bandwidth, and pattern bandwidth. Based on the previous works discussed above [8, 9], the advantages of multiple resonant planar antennas are illustrated and discussed from the following aspects:

1. Planar antennas have attractive physical characteristics, such as low profile, small volume, and light weight. They can be widely used on

missiles, aircraft, automobiles, human bodies, and other conformal structure surfaces while maintaining primary mechanical structures, and flight performances.

2. They can be mass produced at low cost, which simplifies manufacturing and helps facilitate integration with active devices and microwave circuits.
3. They can achieve several desired characteristics, such as linearly polarized (LP) radiation patterns, CP radiation patterns, multiple-polarized radiation patterns, multibandwidth operation, and so on.

At the same time, planar antennas have several disadvantages, such as narrow bandwidth, low-radiation efficiency, and poor polarization purity. All these defects motivate researchers to pursue theoretical investigations and performance improvements, which inspires more research. Here are some main research works conducted during the past three decades.

First, to address the narrow bandwidth for the traditional MPA, Professor K. -M. Luk from the City University of Hong Kong proposed and developed a broadband technique based on the classical *L*-shaped feeding in 1998 [11]. Second, to improve low-radiation gain, Professor D. R. Jackson from the University of California proposed a dielectric-loaded technique on the top surface of the antennas in 1984 [12], which was developed by Professor Zhu [13]. Third, to overcome the high cross-polarization of the MPA at H-plane pattern, Professor Y. -P. Zhang from Nanyang Technological University in Singapore proposed and developed a differential feeding scheme in 2006, which also helps improve impedance matching under differential systems [14].

Although planar antennas have maintained inherent multimode performance, they still have many critical issues that need to be addressed, which we discuss in Chapters 3 to 6. In the following, we would like to illustrate fundamental knowledge for the bandwidth of mode-inspired planar antennas, including impedance bandwidth, isolation bandwidth, gain bandwidth, and pattern bandwidth.

1.3 Relative Bandwidth of Mode-Inspired Planar Antennas

Antenna bandwidth is defined as the range of frequencies within which the performance of the antenna conforms to a specified standard with respect to some characteristic. To simplify the antenna parameter, we use a traditional low-profile MPA as an example to illustrate its input impedance bandwidth, isolation bandwidth, stable far-zone-radiation gain bandwidth, stable radiation beamwidth, stable low-cross-polarization bandwidth, and so on.

1.3.1 Impedance Bandwidth

Initially, the impedance bandwidth of MPA is discussed, and Figure 1.5 plots the input-resistance variation of the antennas. For a traditional MPA in Figure 1.5(a), there are four modes in the wide working band, which are distributed at different resonant frequencies. For the mode-inspired wideband antenna in Figure 1.5(b), the peak input resistances of these four modes are moved in proximity to each other, increasing impedance bandwidth significantly. For the mode-inspired wideband antenna in Figure 1.5(c), the peak input resistances of two modes are moved in proximity to each other, increasing impedance bandwidth under dual working bands. Next, let's study the input impedance calculation method to control the resonant frequencies of these desired modes.

Figures 1.6, 1.7, and 1.8 plot three different types of feeding schemes for the rectangular MPA, which include the aperture-coupled feeding, coaxial probe feeding, and microstrip feeding line. To calculate their input impedance, the equivalent circuits have also been built in Figures 1.6, 1.7, and 1.8. This figure shows that radiating edges are represented as the equivalent RLC (where *R* is the resistance, *L* is the inductance, and *C* is capacitance), and the radiation patch is represented as the transmission line with radiative conductance at two edges. As a result, input impedance Z_{in} is defined as the impedance presented by an antenna at its terminals [9, 19], which could be expressed as

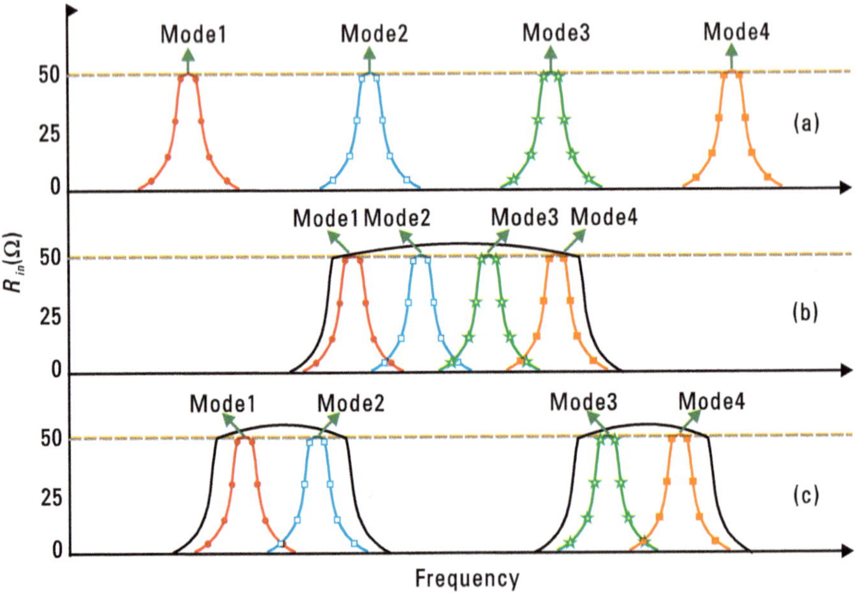

Figure 1.5 Scheme of resonant frequency variation for planar antennas under multimode resonance: (a) traditional antenna, (b) mode-inspired wideband antenna, and (c) mode-inspired multiband antenna.

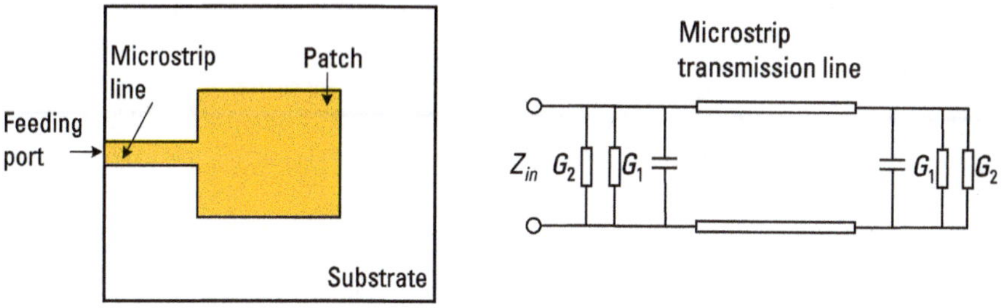

Figure 1.6 Top view and equivalent circuit of a microstrip line-feeding patch antenna [15].

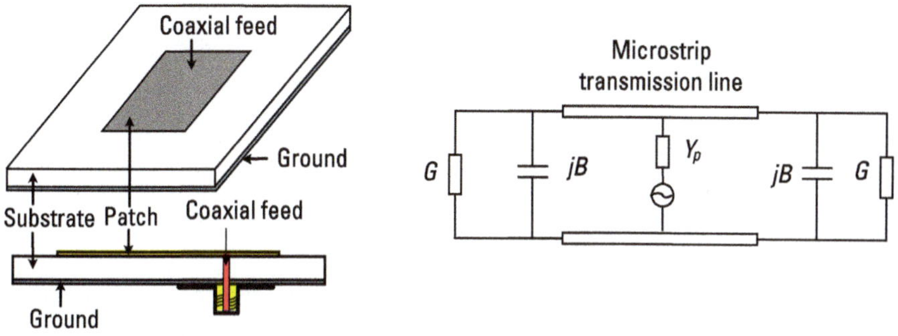

Figure 1.7 Configuration and equivalent circuit of a coaxial feeding patch antenna [16].

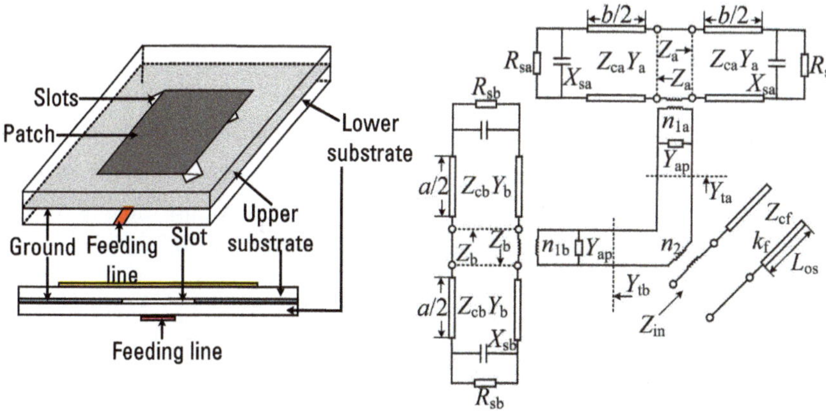

Figure 1.8 Configuration and equivalent circuit of aperture-coupled patch antennas [17, 18].

$$Z_{in} = R_{in} + jX_{in} \tag{1.1}$$

where R_{in} and X_{in} are the real and imaged parts of the input impedance.

Commonly, the input impedance of the antenna needs to be transformed to about 50Ω; otherwise there will be an impedance mismatch at the feeding port. In this condition, we must consider signal reflections, which means that some of the input signals fed to the MPA may be unexpectedly reflected back to the signal sources. The reflection coefficient Γ is introduced and connected with the input impedance by the following equations:

$$\Gamma = \frac{Z_{in} - Z_0}{Z_{in} + Z_0} \tag{1.2}$$

where Z_0 is a source impedance, which is usually the characteristic impedance of the feeding transmission line.

Based on (1.2), the return loss in dB could be defined as

$$RL = -20\log|\Gamma| \tag{1.3}$$

For the majority of well-designed antennas, we use a return loss above 10 dB as the standard. For base station antennas, return loss should usually be kept at least 15 dB, but for mobile terminals, return loss often needs to be maintained at about 6 dB. Alternatively, the voltage standing wave ratio (VSWR) is defined as the magnitude ratio between the maximum voltage V_{max} and minimum voltage V_{min} along the transmission line for characterization of impedance matching or reflection performances, which can be expressed as

$$VSWR = \frac{|V_{max}|}{|V_{min}|} = \frac{1+|\Gamma|}{1-|\Gamma|} \tag{1.4}$$

Based on the above (1.3) and (1.4), the designed antenna is operated within a specified frequency range under a prescribed constant value of VSWR or reflection coefficient. For example, the working band of the antenna within VSWR < 2 is ranged from f_1 to f_2, where f_1 and f_2 stand for the lower and upper edges of the achieved frequency band. The impedance bandwidth of the antenna could be calculated as

$$BW = \frac{f_2 - f_1}{f_0} \times 100\% \tag{1.5}$$

where f_0 is the center frequency of the working band.

With respect to a traditional MPA, impedance bandwidth for VSWR < 2 is kept at around 5% with the thickness of around $0.03\lambda_0$ under single fundamental-mode resonance. Several effective methods of increasing impedance bandwidth are discussed in the following, and the corresponding diagrams are presented in Figures 1.9 to 1.12.

For a traditional MPA in Figure 1.9, only a single mode can be generated under the single-mode operation; the antenna suffers from a narrow impedance bandwidth as exhibited. But for a nonradiative mode-inspired antenna in Figure 1.10, the nonradiative modes from the feeding scheme could be introduced by using the L-shaped probe [11], M-shaped probe [20], and so on, exciting the L-C resonance near the primary mode and forcing the image part of its impedance two times across the zero line as desired. The circular loop could be formed around the center of the Smith chart for impedance bandwidth extension compared to the traditional counterpart in Figure 1.9.

The stacked technique [21, 22] is a proven, powerful tool for impedance bandwidth enhancement. Compared to the traditional MPA in Figure 1.9, two fundamental modes from two separated patches are successfully excited and kept in proximity to each other in Figure 1.11. The impedance bandwidth of the antenna is obviously enhanced with a circular loop on the Smith chart. Due to input resistance with two maximum values, the working band of the antenna is relatively wider than the previous method in Figure 1.9 under the same condition.

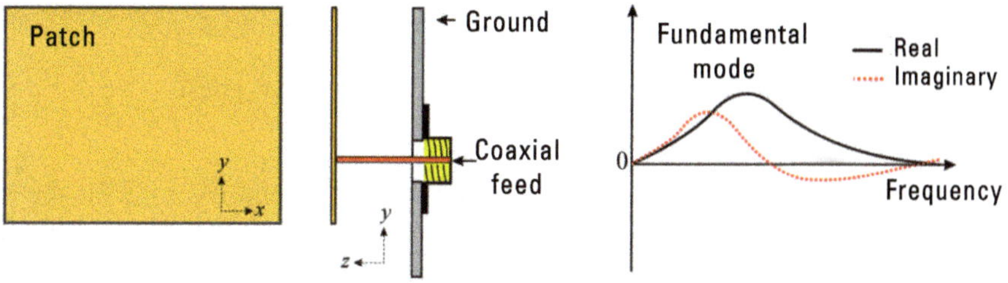

Figure 1.9 Configuration and input impedance of a traditional MPA.

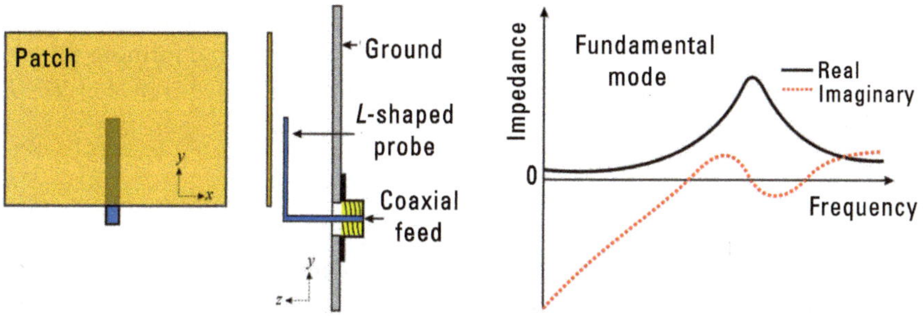

Figure 1.10 Configuration and input impedance of L-probe feeding MPA [11].

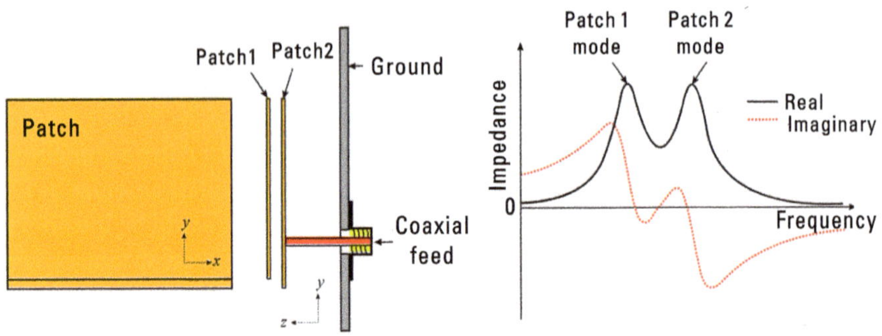

Figure 1.11 Configuration and input impedance of a stacked MPA [21].

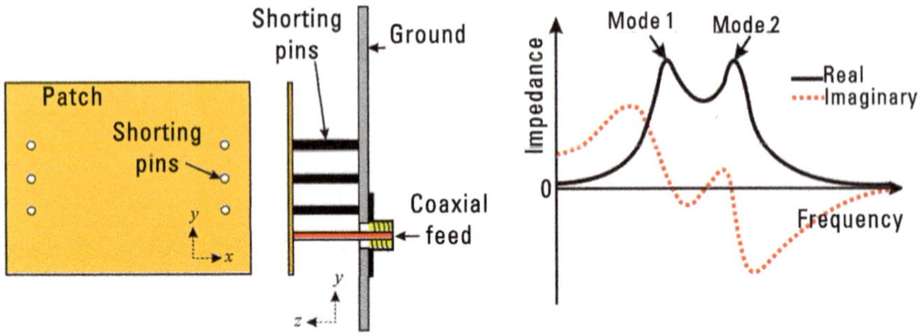

Figure 1.12 Configuration and input impedance of multimode MPAs [7, 24].

Fundamental and higher-order modes are combined [7, 23, 24], reallocating several radiation resistances in proximity to each other. Compared to the traditional MPA in Figure 1.9, the mode 1 and mode 2 from a single radiator could be excited to resonate closely in Figure 1.12; it also generates a circular loop on the Smith chart as desired. Different from the previous stacked method in Figure 1.11, the impedance bandwidth is enhanced by using a single radiator. In other words, if we introduce another patch above the primary patch, four radiative modes can be simultaneously gained for the antenna, forming three or four circular loops on the Smith chart. This method could also be applied for multiband generation in [25].

1.3.2 Isolation Bandwidth

As Figure 1.13 shows, input signals for multiport antennas and arrays are transformed into far-zone-radiated fields or reflection signals themselves, and into other feeding ports to bring mutual coupling or isolation issues. To analyze this

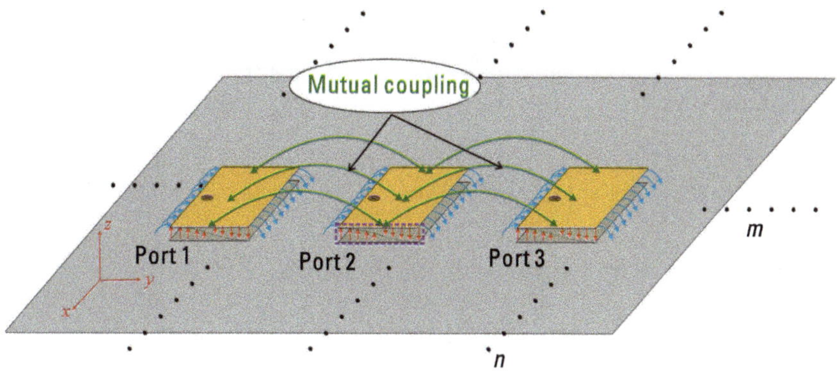

Figure 1.13 Mutual coupling of an array antenna with multifeeding ports.

performance, the scattering matrix is introduced for the antenna to illustrate the mutual coupling or isolation performance.

Taking the S-parameters of an n-port passive network as an example, Figure 1.14 shows the role of S-parameters in the mutual coupling of multiport antennas. Here, the number n of the feeding ports is employed to excite the antenna. For the special feeding port m, its a_m and b_m represent the normalized forward wave and backward wave, respectively.

Based on the classical works [26], the forward wave and backward wave could be connected with the relationship in the following:

$$\begin{cases} b_1 = S_{11}a_1 + S_{12}a_2 + \ldots + S_{1n}a_n \\ b_2 = S_{21}a_1 + S_{22}a_2 + \ldots + S_{2n}a_n \\ \vdots \\ b_n = S_{n1}a_1 + S_{n2}a_2 + \ldots + S_{nn}a_n \end{cases} \tag{1.6}$$

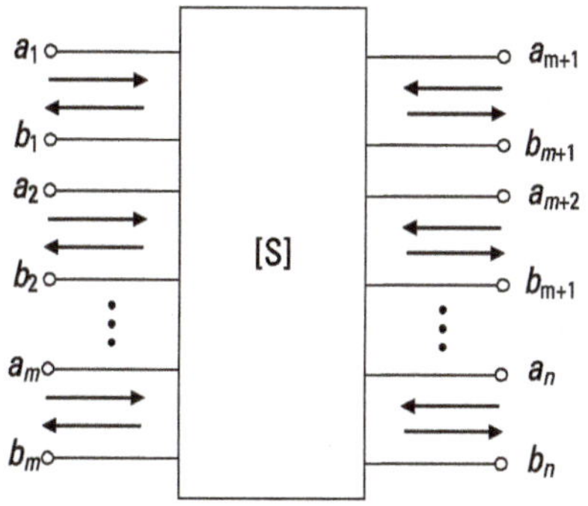

Figure 1.14 Scattering matrix for an n-port network from a multiport antenna.

By using (1.6), the scattering matrix [S] could be formed as

$$\begin{bmatrix} b_1 \\ b_2 \\ \vdots \\ b_n \end{bmatrix} = \begin{bmatrix} S_{11} & S_{12} & \cdots & S_{1n} \\ S_{21} & & & \vdots \\ \vdots & & & \vdots \\ S_{n1} & \cdots & \cdots & S_{nn} \end{bmatrix} \begin{bmatrix} a_1 \\ a_2 \\ \vdots \\ a_n \end{bmatrix} \tag{1.7}$$

Hence, it can be represented in vector form as

$$[b] = [S][a] \tag{1.8}$$

For an n-element antenna array, S_{ii} represents the reflection coefficients under the feeding port i, and S_{ij} $(i \neq j)$ indicates the degree of mutual coupling between the feeding port i and feeding port j. Noticing that, as the value of S_{ij} becomes smaller, the mutual coupling between the port i and port j will decrease. Then, we take the feeding port 1 as an example about its isolation improvement. If we set a_n $(n \neq 1)$ as zero, (1.7) could be simplified as below:

$$\begin{bmatrix} b_1 \\ b_2 \\ \vdots \\ b_n \end{bmatrix} = \begin{bmatrix} S_{11} & S_{12} & \cdots & S_{1n} \\ S_{21} & & & \vdots \\ \vdots & & & \vdots \\ S_{n1} & \cdots & \cdots & S_{nn} \end{bmatrix} \begin{bmatrix} a_1 \\ 0 \\ \vdots \\ 0 \end{bmatrix} \tag{1.9}$$

$$b_1 = S_{11} * a_1 + S_{12} * 0 + S_{13} * 0 \cdots + S_{1n} * 0 \tag{1.10}$$

The input signal from feeding port 1 and other ports is suppressed to improve the isolation performance of the antenna at port 1. If the value of a_n $(n \neq 1)$ can be maintained at zero during the wide operating band, the isolation bandwidth of the antenna at port 1 could be extended as desired. This principle could be applied in other feeding ports, which will be further illustrated in Chapters 2 and 4.

To be extended, by using the S-parameter in scattering matrix from (1.7), the active S-parameter of the array antenna at mth element could be calculated as [27]:

$$actS_m = \left(\sum_{n=1}^{N} A_n e^{j\phi_n} S_{mn} \right) \Big/ \left(A_m e^{j\phi_m} \right) \tag{1.11}$$

where A_n and ϕ_n are the excitation amplitude and phase for the *n*th element, respectively, and S_{mn} is the mutual coupling between its *m*th and *n*th elements.

For (1.11), it shows that mutual coupling is the key parameters for the active S-parameter in the array antennas, which affects the radiation gain in the scanning angle. We must reduce the mutual coupling or improve the isolation between these multiports. For example, for multielements, mutual coupling between closely spaced multiantennas was reduced by using the electromagnetic bandgap (EBG) and defected ground structure (DGS) [28, 29]. The decoupling-feeding-network technique was verified as a useful method for the multiport antennas [30, 31]. In addition, for multiport single antennas, their strong mutual coupling could be shifted down based on the mode-inspired method [32, 33].

1.3.3 Gain Bandwidth

Initially, the average power radiated by an antenna (radiated power) based on the *E* and *H* could be written as

$$\begin{aligned} P_{rad} = P_{av} &= \oiint_S \mathbf{W}_{rad} \cdot ds = \oiint_S \mathbf{W}_{av} \cdot \hat{\mathbf{n}}\, da \\ &= \frac{1}{2} \oiint_S \mathrm{Re}\left(\mathbf{E} \times \mathbf{H}^*\right) \cdot ds \end{aligned} \tag{1.12}$$

where **E** and **H** fields represent the peak values, and they should be omitted for root mean square (RMS) values by using the key value 1/2.

In practice, absolute power patterns are usually not desired. The antenna's performance is simulated and measured by using the gain, in terms of the relative power patterns. Currently, 3D patterns cannot be measured, but they can be constructed with a number of 2D cuts.

Then, the antenna's directivity is defined as the ratio of the radiation intensity in a given direction from the antenna to the radiation intensity averaged over all directions. Average radiation intensity is equal to the total power radiated by the antenna divided by 4π. If the direction is not specified, the direction of the maximum radiation intensity is implied [9]. To simplify the definition, the directivity of the antenna could be expressed as

$$D = \frac{U}{U_0} = \frac{4\pi U}{P_{rad}} \tag{1.13}$$

where U is the radiation intensity of the antenna with the unit of W/unit solid angle.

Meanwhile, the maximum directivity D_0 of the antenna is related to its aperture size and aperture efficiency, which could be illustrated as below

$$A_{em} = \frac{\lambda^2}{4\pi} D_0 \tag{1.14}$$

where A_{em} is the antenna's maximum effective aperture.

Traditionally, the antenna keeps the orthogonal polarization components, so we need to define the partial directivity for a fixed polarization in a fixed direction. The radiation intensity corresponding to a given polarization is divided by the total radiation intensity averaged over all directions. With arrangement for partial directivity, the total maximum directivity value D_0 for the orthogonal θ and ϕ components under a spherical coordinate system could be written as

$$D_0 = D_\theta + D_\phi \tag{1.15}$$

while the partial directivities D_θ and D_ϕ are expressed as

$$D_\theta = \frac{4\pi U_\theta}{(P_{rad})_\theta + (P_{rad})_\phi} \tag{1.16}$$

$$D_\phi = \frac{4\pi U_\phi}{(P_{rad})_\theta + (P_{rad})_\phi} \tag{1.17}$$

In fact, the antenna has multiple losses: dielectric, conduction, impedance mismatching, and so on. A set of efficiencies should be considered for the antenna, as plotted in Figure 1.15. This figure shows that antenna efficiencies consist of total efficiency, reflection efficiency, conduction efficiency, and dielectric efficiency.

In general, the relationship between total efficiency and other efficiencies should be expressed as

$$e_0 = e_r e_c e_d \tag{1.18}$$

The reflection efficiency e_r could be replaced by the reflection coefficient. The total efficiency e_0 is also calculated as

$$e_0 = e_r e_{cd} = e_{cd}\left(1 - |\Gamma|^2\right) \tag{1.19}$$

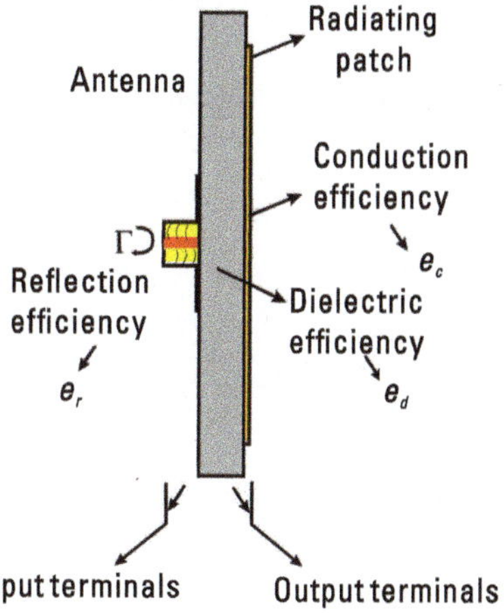

Figure 1.15 Reference terminals and losses for planar antennas.

where $e_{cd} = e_c e_d$ is the radiation efficiency of the antenna, which is adopted to build the relationship between the antenna gain and directivity.

The values of e_c and e_d are difficult to compute, but they can be determined and reduced by properly selecting the thickness of the conductor or dielectric substrate.

We have defined two gains for the antenna, which includes gain with the consideration of radiation efficiency, e_{cd}, and absolute gain with consideration of total efficiency, e_0. Based on (1.19), the gain of the antenna could be written as

$$G(\theta,\phi) = e_{cd} D(\theta,\phi) \tag{1.20}$$

Meanwhile, the absolute gain of the antenna could be written as

$$\begin{aligned} G_{abs}(\theta,\phi) &= e_r G(\theta,\phi) = \left(1-|\Gamma|^2\right) G(\theta,\phi) \\ &= e_r e_{cd} D(\theta,\phi) = e_o D(\theta,\phi) \end{aligned} \tag{1.21}$$

Similar to directivity, we can also define realized gain at special polarization and special direction as "the part of radiation intensity corresponding to a given polarization divided by the total radiation intensity that would be obtained if the power accepted by the antenna were radiated isotopically." With this arrangement, the maximum gain G_0 of the antenna could be decomposed into the orthogonal θ and ϕ components, which are expressed as

$$G_0 = G_\theta + G_\phi \tag{1.22}$$

where the partial gains G_θ and G_ϕ are written as

$$G_\theta = \frac{4\pi U_\theta}{P_{in}} \tag{1.23}$$

$$G_\phi = \frac{4\pi U_\phi}{P_{in}} \tag{1.24}$$

where P_{in} is the total input power or accepted power.

In addition, from (1.14), the maximum directivity of the antenna is mainly determined by the aperture size and aperture efficiency. The maximum absolute gain of the antenna could be written as below

$$G_{0abs} = e_r e_{cd} \left. D(\theta, \varphi) \right|_{\max} \tag{1.25}$$

$$\left. D(\theta, \varphi) \right|_{\max} = A_{em} \frac{4\pi}{\lambda^2} \tag{1.26}$$

where A_{em} is the total input power or accepted power, which is about the physical area of the antenna multiplied by its aperture efficiency.

In (1.25) and (1.26) it shows that the realized gain of an antenna is influenced by key factors such as physical area, aperture efficiency, the impedance mismatch factor e_r, conduction efficiency e_c, and dielectric efficiency e_d. We would like to illustrate the gain-enhancement method in more in detail.

First, antenna losses should be entirely reduced for gain enhancement, such as reflection (mismatch) loss, conduction loss, and dielectric loss. For reflection (mismatch) loss, an antenna design must have a minimum VSWR. For example, the VSWR of the horn antennas is usually kept below 1.2 or even 1.1, and the VSWR of the base station antennas is usually kept below 1.5. For conduction loss, conductor thickness should be minimized while keeping conductivity high with silver. For dielectric loss, the loss tangent of the dielectric substrate should be kept as small as possible; an air substrate could be introduced to reduce dielectric loss. As reported in [23], selecting substrate FR4 with high dielectric loss can reduce radiation gain by 1.5 dBi or even further.

Second, the antenna's physical area could be enlarged for gain enhancement. For example, a dielectric substrate could be suspended on the antenna's top layer [12, 13], increasing the total aperture size of the antenna and achieving a high gain of above 15 dB. Meanwhile, an array antenna with multielements could increase aperture size to achieve high gain of about 14 dB [34].

Third, antenna aperture efficiency could be enhanced for gain improvement. For example, the electric-field distributions of the antenna in [35] are kept more uniform around the edges of the radiating patch by using a slot. Its radiation gain enhancement can exceed 3 dB without an enlarged electric size. As reported in [36, 37], the aperture efficiency of an array antenna is enhanced by using phoenix elements or horn antennas, leading to a high gain of above 25 dB.

If we try to improve gain bandwidth, antenna gain in the wideband or multiband should be maintained as stable as possible. But with increasing resonant frequency, the aperture size of the antenna is obviously enlarged. So antenna gains in the upper band are always higher than those in the lower band. To keep stable characteristics, the most effective approach is to control the concerned aperture size or aperture efficiency during the wideband or multiband.

1.3.4 Pattern Bandwidth

A radiation pattern is defined as the spatial distribution of a quantity that characterizes the electromagnetic field generated by an antenna. This pattern is represented and calculated as a function of the directional coordinates, which include power flux density, radiation intensity, field strength, directivity, phase or polarization, and so on.

Antennas can have many kinds of radiation patterns. First, an idealized isotropic radiator, which has equal radiation in all directions as shown in Figure 1.16(a), would be in high demand for wireless communications, but it cannot be physically realized. As a result, the isotropic radiator is often taken as a reference for quantitative expression of the directive properties of practical antennas. Second, the bidirectional radiation pattern derived from the slot antenna is plotted in Figure 1.16(b) at the first-order mode. Third, the omnidirectional pattern from a dipole antenna at the first-order mode is considered as a special type of directional pattern, which keeps the maximum gain at one plane, as shown in Figure 1.16(c). Fourth, radiation patterns from a patch antenna with a directional beam in the special direction are shown in Figure 1.16(d), which could be well applied in point-to-point communications, radar systems, and so on. To simplify the discussion of radiation patterns, we would like to take the directional beam in Figure 1.16(d) as an example in the following [9].

Figure 1.17 plots an amplitude field pattern based on the tracing of a received electric (magnetic) field at a constant radius. Usually, the field pattern is normalized to its maximum value, yielding normalized field and power patterns. Meanwhile, the field pattern is always plotted on a logarithmic scale or more commonly in decibels. This figure shows that the field pattern contains

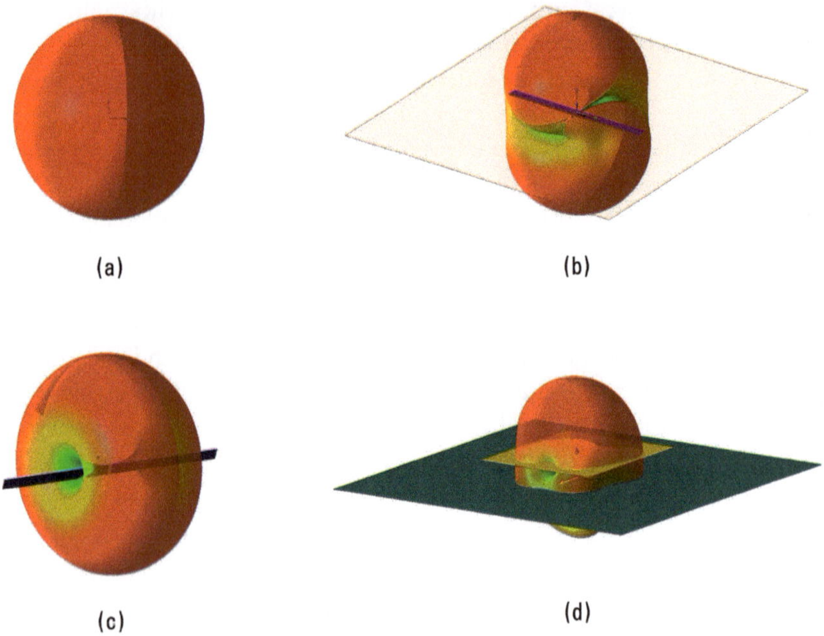

Figure 1.16 Four different types of 3D-view radiation patterns from different antennas: (a) ideally isotropic pattern, (b) bidirectional pattern from slot antennas, (c) omnidirectional pattern from dipole antennas, and (d) directional pattern from patch antennas [9].

the main beam, sidelobe, backlobe, and so on, which keep their maximum value at different angles as a function of the angular space.

In practical applications, a 3D-radiation pattern is not commonly used to analyze antenna performance because it lacks polarization information. As an alternative, 2D normalized field patterns are adopted as a function of θ for some particular values of ϕ, plus a few plots as a function of ϕ for some particular values of θ, herein the field pattern of the linear antenna array, which is plotted under the linear scale in Figure 1.18. The field pattern contains the main beam, sidelobe, backlobe, and so on. Meanwhile, plus (+) and minus (-) are signed around the main beam and sidelobes to indicate the relative polarization (positive or negative) of the amplitude between the various lobes. Field nulls are formed across their variation.

The main lobe (also called main beam) is defined as the radiation lobe containing the direction of maximum radiation. As observed in Figure 1.18, the major lobe of an array antenna points to $\theta = 0$. Some special applications, such as dual-beam antennas, have two major beams off-broadside direction. The main beam represents radiation in desired directions, and should be kept with the maximum value.

A sidelobe is defined as the radiation lobe in any direction other than that of the major lobe. Except for the main beam, all lobes could be called sidelobes.

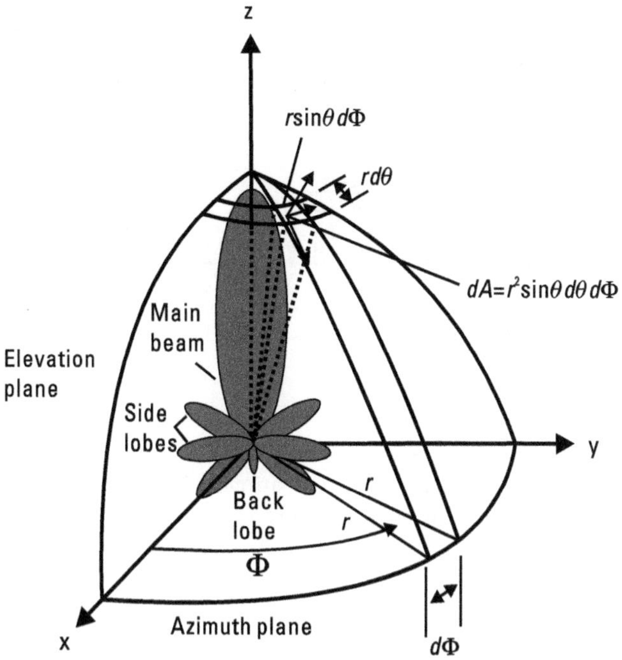

Figure 1.17 Coordinate system for the antenna with directional radiation patterns [9].

Usually, the mainly concerned sidelobe of the antenna is adjacent to the main beam and occupies the hemisphere in the direction of the main beam. Sidelobes represent radiation in undesired directions, and they should be minimized or suppressed.

A backlobe is defined as a radiation lobe with an axis approximately 180° off the beam axis of an antenna lobe in the half-space opposed to the direction of peak directivity. It traditionally refers to a minor lobe that occupies the hemisphere in a direction opposite that of the main beam. The backlobe represents radiation in undesired directions, which should also be minimized.

Based on [9], for the main beam, the beamwidth shown in Figure 1.18, is the key antenna parameter, defined as the angular separation between two identical points on opposite sides of the pattern maximum. In an antenna pattern, a large number of beamwidth properties are performed. One of the most widely used beamwidths of the antenna is named as the half-power beamwidth (HPBW), which is defined this way by IEEE: "In a radiation-pattern cut containing the direction of the maximum of a lobe, the angle between the two directions in which the radiation intensity is one-half the maximum value." Another important antenna beamwidth is the angular separation between the first nulls of far-zone-radiation patterns, called first-null beamwidth (FNBW). Both HPBW and FNBW are given in Figure 1.18. The beamwidth of an antenna is

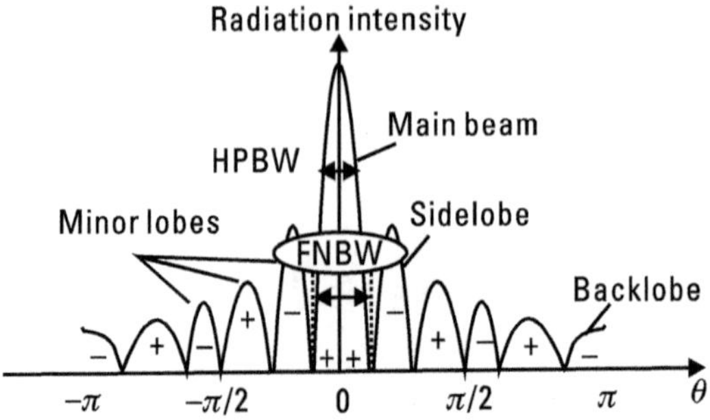

Figure 1.18 Normalized electric-field pattern for a linear array with a linear scale.

an important figure of merit and is used as a trade-off between its main beam and sidelobe level. Usually, the beamwidth of the antenna is employed to describe the resolution capabilities of the antenna to distinguish between two adjacent radiating sources or radar targets.

For a LP antenna, performance is often described based on two principal planes, inclusive of *E*-plane pattern and *H*-plane pattern. The *E*-plane is defined as the plane containing the *electric*-field vector and the direction of maximum radiation. The *H*-plane is defined as the plane containing the *magnetic*-field vector and the direction of maximum radiation. For example, a slot antenna with the bidirectional pattern in Figure 1.19 has an infinite number of principal *H*-planes, and one principal *E*-plane at the first-order mode. A dipole antenna with the omnidirectional pattern in Figure 1.20 has an infinite number of principal *H*-planes, and one principal *E*-plane at the first-order mode. An illustration of the MPA at TM_{10} mode is shown in Figure 1.21, the *x*-*z* plane ($\phi = 0°$) is its principal *E*-plane and the *y*-*z* plane ($\phi = 90°$) is its principal *H*-plane.

The HPBW of an antenna can be divided into the *E*-plane HPBW and *H*-plane HPBW. In the following, we hope to illustrate the beamwidth-enhancement method more in detail.

First, for the *E*-plane HPBW, the radiation beam of a single MPA is mainly controlled by the parallel equivalent-magnitude currents at patch edges. If we try to increase its HPBW, the electrical distance of its equivalent-magnitude currents should be reduced down [38], which could be realized by controlling the resonant frequency of its fundamental or higher-order modes with the fixed total dimensions. For an array antenna, the *E*-plane HPBW is also controlled by setting the spacing between different elements. If we try to increase the HPBW, the distance at *E*-plane between the adjacent elements should be kept as small as possible [9]. However, this method may lead to serious mutual coupling between its multiport, reducing the isolation bandwidth of the array antenna.

Second, for the H-plane HPBW, the radiation beam of a single MPA is mainly determined by the width of the radiating patch. If we try to increase its

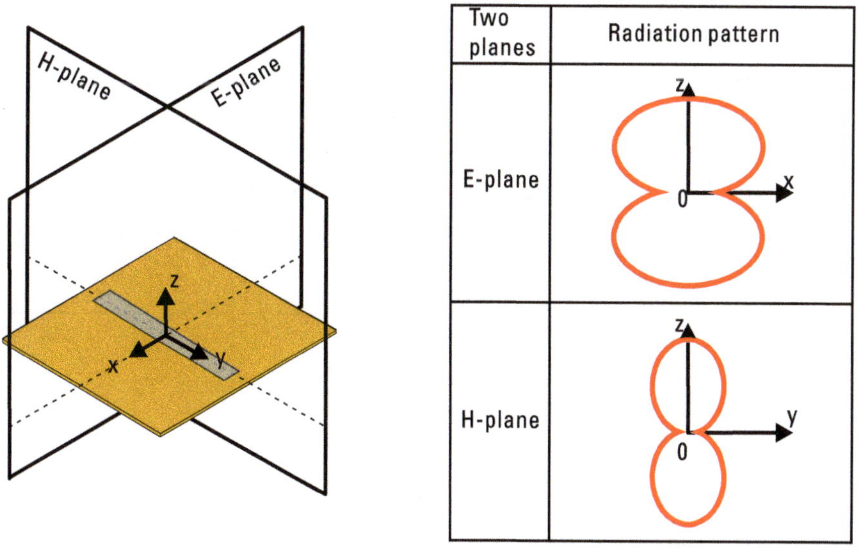

Figure 1.19 Configuration, *E*-plane radiation pattern, and *H*-plane radiation pattern of a slot antenna.

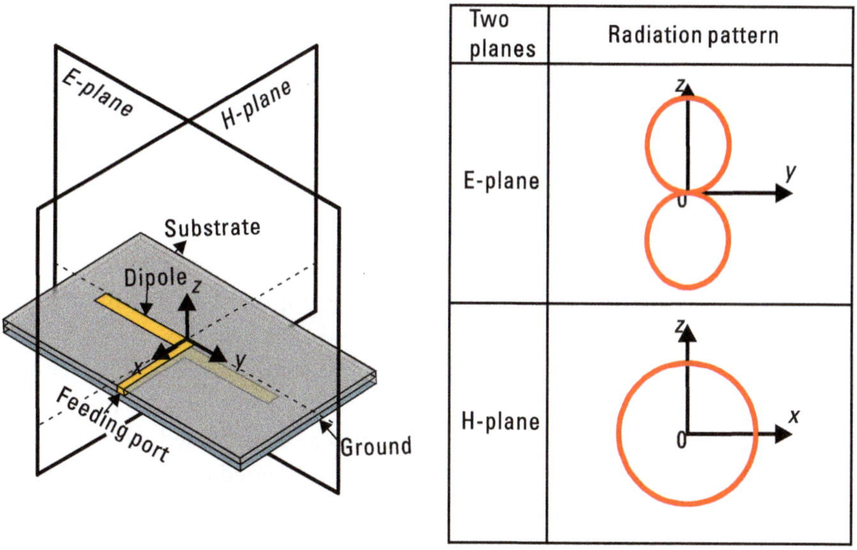

Figure 1.20 Configuration, *E*-plane radiation pattern, and *H*-plane radiation pattern of a printed dipole antenna.

HPBW, the electrical width of the patch should be reduced, which could be realized by electrically shorting the radiating edges partly [6]. Nevertheless, this method leads to small gain. Additional radiated fields could be introduced and

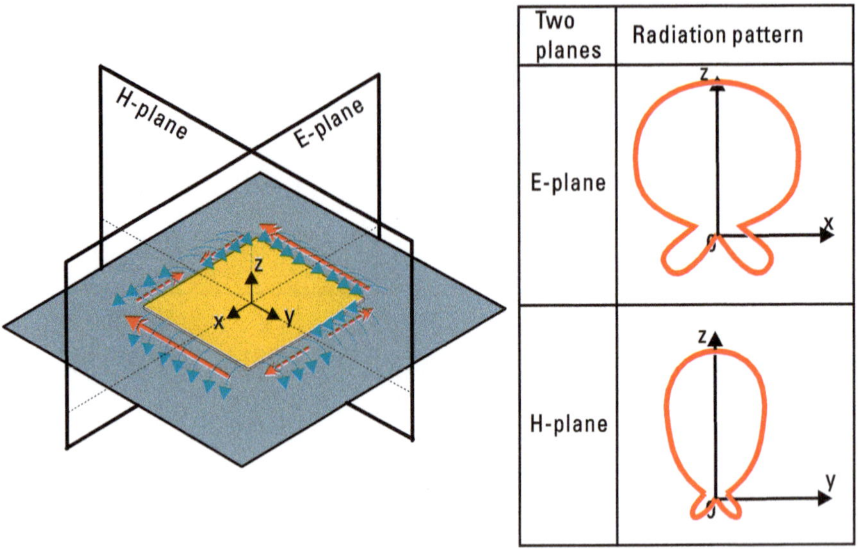

Figure 1.21 Configuration, *E*-plane radiation pattern, and *H*-plane radiation pattern of an MPA at TM_{10} mode.

combined with the primary fields, increasing gain off-broadside direction [39, 40]. Unfortunately, the HPBW of a low-profile printed antenna at H-plane is usually increased to about 120°. The H-plane HPBW could be enhanced by folding the ground plane [41]. The H-plane HPBW of an array antenna is also controlled by the spacing between different elements. If we try to increase its value, the distance of the antenna element at H-plane should be kept as small as possible [9]. However, this method will lead to serious mutual coupling between its multiport, reducing the isolation bandwidth of the array antenna.

Third, if the HPBW at these two planes is improved for the antenna, we can explore this method for CP-beamwidth enhancement. As shown in [42–44], the HPBW of their E_ϕ components could be controlled so as to match well with that of E_θ components, leading to the wide AR-beamwidth in the wide angle. In [45], the vertical patches were loaded and connected with the main radiating patch, aiming to increase the vertically oriented currents and extend the antenna's CP beamwidth.

Antenna pattern bandwidth is highly demanded in wireless communication, such as the stable HPBW of about 65° for base station antenna. If we try to improve this value, the radiation pattern of the antenna in the wideband or multiband should be maintained to the extent possible. Unfortunately, with increasing resonant frequency, the antenna HPBW is reduced. So the HPBW of the antenna in the upper band is always narrower than that of the lower band. The most effective way to address this issue is to excite different radiation beams under different elements [46]. Meanwhile, the electric-field distributions of the antenna could be controlled so as to realize this performance as desired.

1.4 Concluding Remarks

This chapter introduced the fundamental works of several planar antenna types: slot antenna, printed dipole antenna, and MPA. Next, the MPA is taken as a representative example for antenna parameter discussion, such as input impedance response, isolation bandwidth, radiation gain, radiation pattern, beamwidth, sidelobe level, and so on. In addition, we briefly discussed a few effective methods for antenna bandwidth enhancement. These materials will provide theoretical and methodological guidance for the wideband or multiband design of impedance bandwidth in Chapter 3, isolation bandwidth in Chapter 4, gain bandwidth in Chapter 5, and pattern bandwidth in Chapter 6.

References

[1] Garg, R., et al., *Microstrip Antenna Design Handbook*, Norwood, MA: Artech House, 2001.

[2] Zhu, L., R. Fu, and K. -L. Wu, "A Novel Broadband Microstrip-Fed Wide Slot Antenna with Double Rejection Zeros," *IEEE Antennas and Wireless Propagation Letters,* Vol. 2, 2003, pp. 194–196.

[3] Lu, W. -J., and L. Zhu, "Wideband Stub-Loaded Slotline Antennas Under Multimode Resonance Operation," *IEEE Transactions on Antennas and Propagation,* Vol. 63, No. 2, Feb. 2015, pp. 818–823.

[4] Zhu, L., and K. Wu, "Model-Based Characterization of CPS-Fed Printed Dipole for Innovative Design of Uniplanar Integrated Antenna," *IEEE Microwave and Guided Wave Letters,* Vol. 9, No. 9, Sep. 1999, pp. 342–344.

[5] Zhang, Y., and L. Zhu, "Printed Dual Spiral-Loop Wire Antenna for Broadband Circular Polarization," *IEEE Antennas and Wireless Propagation Letters*, Vol. 54, No. 1, Jan. 2006, pp. 284–288.

[6] Liu, C. -M., S. Xiao, and X. -L. Zhang, "A Compact, Low-Profile Wire Antenna Applied to Wide-Angle Scanning Phased Array," *IEEE Antennas and Wireless Propagation Letters,* Vol. 17, No. 3, March 2018, pp. 389–392.

[7] Liu, J., and Q. Xue, "Broadband Long Rectangular Patch Antenna with High Gain and Vertical Polarization," *IEEE Transactions on Antennas and Propagation,* Vol. 61, No. 2, February 2013, pp. 539–546.

[8] Zhu, L., S. Sun, and W. Menzel, "Ultra-Wideband (UWB) Bandpass Filters Using Multiple-Mode Resonator," *IEEE Microwave and Wireless Components Letters,* Vol. 15, No. 11, November 2005, pp. 796–798.

[9] Balanis, Constantine A., *Antenna Theory Analysis and Design*, Hoboken, NJ: Wiley, 2005.

[10] Lu, W. -Ju, and L. Zhu, *Multi-Mode Resonant Antennas: Theory, Design, and Applications,* Boca Raton, FL: CRC Press, 2022.

[11] Luk, K. -M., et al., "Broadband Microstrip Patch Antenna," *IEEE Electron Device Lett.*, Vol. 34, No. 15, July 1998, pp. 1442–1443.

[12] Alexopoulos, N., and D. Jackson, "Fundamental Superstrate (Cover) Effects on Printed Circuit Antennas," *IEEE Transactions on Antennas and Propagation,* Vol. 32, No. 8, August 1984, pp. 807–816.

[13] Zhu, L., I. Joishi, and E. Yamashita, "Characterization of Microstrip Patch Antennas Suspended by a Dielectric Superstrate with High Permittivity," *IEEE Antennas and Propagation Society International Symposium,* 1996 Digest, Vol. 1, Baltimore, MD, 1996, pp. 704–707.

[14] Zhang, Y. P., and J. J. Wang, "Theory and Analysis of Differentially-Driven Microstrip Antennas," *IEEE Transactions on Antennas and Propagation,* Vol. 54, No. 4, April 2006, pp. 1092–1099.

[15] Ledee, R., et al., "An Extension of the Transmission Line Model for the Characterization of Microstrip-Microslot Applicator," *18th European Microwave Conference,* Stockholm, Sweden, 1988, pp. 588–592.

[16] Schaubert, D., et al., "Microstrip Antennas with Frequency Agility and Polarization Diversity," *IEEE Transactions on Antennas and Propagation,* Vol. 29, No. 1, January 1981, pp. 118–123.

[17] Vlasits, T., E. Korolltiewicz, and A. Sambell, "Analysis of Cross-Aperture Coupled Patch Antenna Using Transmission Line Model," *Electronics Letters,* Vol. 32, No. 21, October 1996, pp. 1934–1935.

[18] Al Jibouri, B., et al., "Cavity Model of Circularly Polarised Cross-Aperture-Coupled Microstrip Antenna," *IEE Proceedings-Microwaves, Antennas and Propagation,* Vol. 148, No. 3, June 2001, pp. 147–152.

[19] "IEEE Standard for Definitions of Terms for Antennas," *IEEE Std (2014)* 145–2013 (*Revision of IEEE Std 145–1993*), pp. 1–50.

[20] Lin, Q. W., et al., "Printed Meandering Probe-Fed Circularly Polarized Patch Antenna with Wide Bandwidth," *IEEE Antennas and Wireless Propagation Letters,* Vol. 13, 2014, pp. 654–657.

[21] Croq, F., and A. Papiernik, "Stacked Slot-Coupled Printed Antenna," *IEEE Microwave and Guided Wave Letters,* Vol. 1, No. 10, October 1991, pp. 288–290.

[22] Liu, S., W. Wu, and D. -G. Fang, "Single-Feed Dual-Layer Dual-Band E-Shaped and U-Slot Patch Antenna for Wireless Communication Application," *IEEE Antennas and Wireless Propagation Letters,* Vol. 15, 2016, pp. 468–471.

[23] Xiao, S., et al., "Bandwidth-Enhancing Ultralow-Profile Compact Patch Antenna," *IEEE Transactions on Antennas and Propagation,* Vol. 53, No. 11, Nov. 2005, pp. 3443–3447.

[24] Wong, H., K. K. So, and X. Gao, "Bandwidth Enhancement of a Monopolar Patch Antenna with V-Shaped Slot for Car-to-Car and WLAN Communications," *IEEE Transactions on Vehicular Technology,* Vol. 65, No. 3, Mar. 2016, pp. 1130–1136.

[25] Zhong, S. S., and Y. T. Lo, "Single-Element Rectangular Microstrip Antenna for Dual-Frequency Operation," *IEEE Electronics Letters,* Vol. 19, No. 8, Apr. 1983, pp. 298–300.

[26] Zhu, L., S. Sun, and R. Li, *Microwave Bandpass Filters for Wideband Communications,* Hoboken, NJ: Wiley, 2012.

[27] Kalfa, M., and E. Halavut, "A Fast Method for Obtaining Active S-Parameters in Large Uniform Phased Array Antennas," *2013 IEEE International Symposium on Phased Array Systems and Technology,* Waltham, MA, 2013.

[28] Chiu, C. -Y., et al., "Reduction of Mutual Coupling Between Closely-Packed Antenna Elements," *IEEE Transactions on Antennas and Propagation,* Vol. 55, No. 6, June 2007, pp. 1732–1738.

[29] Farahani, H. S., et al., "Mutual Coupling Reduction in Patch Antenna Arrays Using a UC-EBG Superstrate," *IEEE Antennas and Wireless Propagation Letters,* Vol. 9, 2010, pp. 57–59.

[30] Zhao, L., and K. -L. Wu, "A Dual-Band Coupled Resonator Decoupling Network for Two Coupled Antennas," *IEEE Transactions on Antennas and Propagation,* Vol. 63, No. 7, July 2015, pp. 2843–2850.

[31] Sui, J., and K. -L. Wu, "A General T-Stub Circuit for Decoupling of Two Dual-Band Antennas," *IEEE Transactions on Microwave Theory and Techniques,* Vol. 65, No. 6, June 2017, pp. 2111–2121.

[32] Liu, N. -W., et al., "Electric-Field Null Bending of a Single Dual-Port Patch Antenna for Colinear Polarization Decoupling Using Characteristic Modes Analysis," *IEEE Transactions on Antennas and Propagation,* Vol. 70, No. 12, December 2022, pp. 12247–12252.

[33] Liu, N. -W., et al., "Low-Profile Four-Port Patch Antenna with Wide Isolation-Bandwidth and Same Polarization: Principle and Design Approach," *IEEE Antennas and Wireless Propagation Letters,* doi: 10.1109/LAWP.2023.3289575.

[34] Yang, G., et al., "Study on Wide-Angle Scanning Linear Phased Array Antenna," *IEEE Transactions on Antennas and Propagation,* Vol. 66, No. 1, January 2018, pp. 450–455.

[35] Zhang, X., and L. Zhu, "Gain-Enhanced Patch Antennas with Loading of Shorting Pins," *IEEE Transactions on Antennas and Propagation,* Vol. 64, No. 8, August 2016, pp. 3310–3318.

[36] Deng, R., et al., "Single-Layer Dual-Band Reflect Array Antennas with Wide Frequency Ratios and High Aperture Efficiencies Using Phoenix Elements," *IEEE Transactions on Antennas and Propagation,* Vol. 65, No. 2, February 2017, pp. 612–622.

[37] Bhattacharyya, A. K., and G. Goyette, "A Novel Horn Radiator with High Aperture Efficiency and Low Cross-Polarization and Applications in Arrays and Multibeam Reflector Antennas," *IEEE Transactions on Antennas and Propagation,* Vol. 52, No. 11, November 2004, pp. 2850–2859.

[38] Liu, N. -W., et al., "Low-Profile Microstrip Patch Antenna with Simultaneous Enhanced Bandwidth, Beamwidth, and Cross-Polarization Under Dual-Resonance," *IET Microwaves, Antennas & Propagation,* Vol. 14, No. 5, April 2020, pp. 360–365.

[39] Chen, Z. N., W. K. Toh, and X. Qing, "A Microstrip Patch Antenna with Broadened Beamwidth," *Microwave and Optical Technology Letters,* Vol. 50, No. 7, July 2008, pp. 1885–1888.

[40] Kornprobst, J., et al., "A mm-Wave Patch Antenna with Broad Bandwidth and a Wide Angular Range," *IEEE Transactions on Antennas and Propagation,* Vol. 65, No. 8, August 2017, pp. 4293–4298.

[41] Li, Y., and K. -M. Luk, "A Linearly Polarized Magnetoelectric Dipole with Wide H-Plane Beamwidth," *IEEE Transactions on Antennas and Propagation,* Vol. 62, No. 4, April 2014, pp. 1830–1836.

[42] Luo, Y., Q. -X. Chu, and L. Zhu, "A Low-Profile Wide-Beamwidth Circularly-Polarized Antenna Via Two Pairs of Parallel Dipoles in a Square Contour," *IEEE Transactions on Antennas and Propagation,* Vol. 63, No. 3, March 2015, pp. 931–936.

[43] You, M., et al., "A Novel Planar Endfire Circularly Polarized Antenna with Wide Axial-Ratio Beamwidth and Wide Impedance Bandwidth," *IEEE Transactions on Antennas and Propagation,* Vol. 64, No. 10, October 2016, pp. 4554–4559.

[44] Zhang, X., L. Zhu, and N. -W. Liu, "Pin-Loaded Circularly-Polarized Patch Antennas with Wide 3-dB Axial Ratio Beamwidth," *IEEE Transactions on Antennas and Propagation,* Vol. 65, No. 2, February 2017, pp. 521–528.

[45] Choi, E. -C., J. W. Lee, and T. -K. Lee, "Modified S-Band Satellite Antenna with Isoflux Pattern and Circularly Polarized Wide Beamwidth," *IEEE Antennas and Wireless Propagation Letters,* Vol. 12, 2013, pp. 1319–1322.

[46] Chu, Q. -X., and Y. Luo, "A Broadband Unidirectional Multi-Dipole Antenna with very Stable Beamwidth," *IEEE Transactions on Antennas and Propagation,* Vol. 61, No. 5, May 2013, pp. 2847–2852.

2

Multimode Resonance-Inspired Theory of Wideband and Multiband Planar Antennas

2.1 Introduction

As introduced in Chapter 1, there are a large number of planar antennas: printed slot antennas, MPAs, printed dipole antennas, and so on. With slot antennas, the first mode, second mode, and third mode can be simultaneously excited during wideband use; their electric-field outline proves these modes with different wavelengths. Dipole antennas also excite fundamental and higher-order modes, which keep different wavelengths at different modes. For these two antennas, electric-field distributions are only varied in 1D degree. With MPAs, electric-field distribution varies along the x-direction or the y-direction. MPAs have attractive performance in 2D degrees compared to dipoles and slots. In this chapter, we will use a traditional MPA as an example to theoretically analyze these radiative modes in more detail.

Section 2.2 introduces mode analysis in planar patch antennas based on an equivalent-cavity model. Section 2.3 discusses resonant frequencies of a traditional MPA under multimode resonance. Section 2.4 illustrates the radiation patterns of a traditional MPA under multimode resonance. Section 2.5 shows how the input impedance bandwidth of an antenna under multimode resonance can be improved for wideband or multiband performance. Section 2.6 discusses the isolation bandwidth of a traditional MPA under multimode resonance, enhanced through examination of combined electric-field nodes. Section 2.7 illustrates the radiation gain bandwidth of a traditional MPA under multimode resonance, inclusive of array-factor controlling and additional element methods. Section 2.8 discusses the pattern bandwidth of a traditional MPA under multimode resonance, divided into E-plane and H-plane beamwidth enhancements, plus cross-polarization reduction. Section 2.9 provides a multimode resonance-inspired theory of planar patch antennas with wideband or multiband performance.

2.2 Mode Analysis in Planar Antennas

2.2.1 Cavity Model

To simplify analysis, we will use a traditional MPA as an example. Figure 2.1 plots the detail configuration of a traditional thin, rectangular MPA. It contains a patch radiator on the top layer, a dielectric substrate in the middle layer, and a ground plane on the bottom layer. A feeding probe is inserted in the dielectric substrate to electrically connect the radiating patch and ground plane. Here, *a* and *b* represent the length and width of the radiating patch, respectively, and *h* is the thickness of the dielectric substrate.

As [1] shows, the cavity-model theory should be based on the following observations: First, the electric field inside the cavity keeps invariant along the z-direction ($\partial/\partial_z = 0$), since the thickness *h* is kept much smaller than the

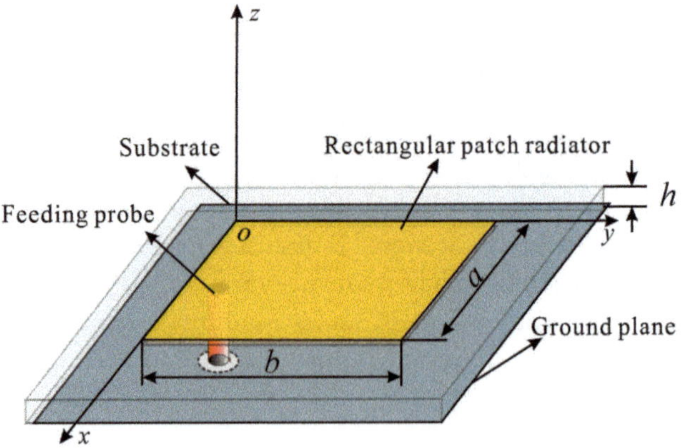

Figure 2.1 Configuration of a traditional rectangular MPA with the feeding probe.

free-space wavelength λ_0. Second, due to the electric walls on the top-layer radiating patch and bottom-layer ground plane, the magnetic field has only transverse components in these places. Third, the electric current on the radiating patch has no component normal to the edge of the patch metallization. The wave numbers k_{mn} and eigenfunctions Ψ_{mn} of the rectangular MPA in Figure 2.1 for nonstatic, resonant TM_{mn} modes are

$$k_{mn} = \sqrt{\left(\frac{m\pi}{a}\right)^2 + \left(\frac{n\pi}{b}\right)^2} \tag{2.1}$$

$$\psi_{mn} = C_{mn} \cos\frac{m\pi x}{a} \cos\frac{n\pi y}{b} \tag{2.2}$$

where C_{mn} is the coefficient TM_{mn} mode, m and n are integers equal to 0,1,2.... that cannot simultaneously be zero, and the thickness h is ignored.

Based on (2.1) and (2.2), which are the linear combination of the eigenfunctions, the internal electric-field distribution of the antenna within the cavity region could be expressed as

$$\begin{aligned} E_z &= \sum_{m,n} A_{mn} \psi_{mn} \\ &= \sum_{m,n} A_{mn} C_{mn} \cos\frac{m\pi x}{a} \cos\frac{n\pi y}{b} \\ &= \sum_{m,n} B_{mn} \cos\frac{m\pi x}{a} \cos\frac{n\pi y}{b} \end{aligned} \tag{2.3}$$

where A_{mn} and B_{mn} in (2.3) are inversely proportional to $(k^2 - k_{mn}{}^2)$. Meanwhile, it can be seen in (2.3) that the antenna has different wave numbers k_{mn} under different resonant frequencies. When the antenna is operated at f_{mn}, its value of k and k_{mn} are maintained in proximity to each other, resulting in a small value of $(k^2 - k_{mn}{}^2)$, which leads to a large value of the internal field amplitude B_{mn}. The internal electric-field variation of the antenna is mainly determined by the B_{mn}. Under this condition, only the TM_{mn} mode is operated for the MPA.

By using (2.3), Figure 2.2 plots the internal electric-field distributions of the MPA under the fundamental and several higher-order modes, such as TM_{01}, TM_{02}, TM_{21}, and TM_{03}. Next, we will discuss the electric-field performance more in detail.

Figure 2.2(a) shows that the internal electric field E_z between the radiating patch and ground plane keeps the same direction along the x-axis, resulting in a value of m as 0. But the internal electric field E_z varies once along the y-axis, which has the half-wavelength property, resulting in a value of n as 1. Due to this arrangement, TM_{01} mode is defined for the antenna. From Figure 2.2(b), the internal electric field E_z still remains constant along the x-axis, so the value of m should be selected as 0, and the internal electric field E_z along the y-axis has full-wavelength performance with two nodes, which indicates that the value of n should be selected as 2. The antenna operates in TM_{02} mode under this condition. Similar to this principle, as shown in Figure 2.2(c), the TM_{21} mode is excited for the two electric-field nodes along the x-direction and one electric-field node along the y-direction. The results in Figure 2.2(d) show that the electric fields under TM_{03} mode resonance remain constant along the x-direction, and possess three nodes along the y-direction.

When boundary conditions of the MPA are damaged (such as the loaded shorting pins, slots, open stubs, and so on), if we continue to name these modes according to the numerical function definition described in (2.3), results may be inaccurate. For example, the theoretical solution of these functions is difficult to calculate, and the mode is difficult to name and track with a confusing

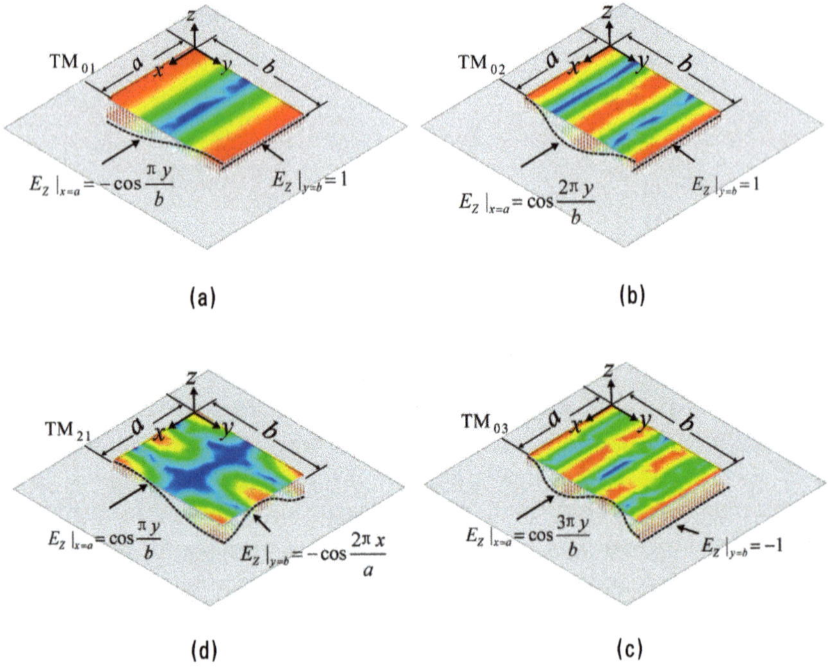

Figure 2.2 Internal electric-field distribution of a traditional rectangular MPA under four different modes: (a) TM_{01} mode, (b) TM_{02} mode, (c) TM_{21} mode, and (d) TM_{03} mode.

number of variations for readers. Therefore, we would like to simplify the mode-tracking process by accounting for the electric-field nodes of these resonant modes, which makes it easier for us to fix the values of *m* and *n*. Except for this method, to avoid confusion, using the characteristic modes [2], such as CM1, CM2, and CM3, is an effective approach to track these modes, since the numbers 1, 2, and 3 are unconnected with the value of *m* and *n* from (2.1) to (2.3).

2.2.2 Even-Order and Odd-Order Modes

Section 2.2.1 uses the cavity model to calculate the internal electric field of different modes. In these radiative modes, there are different electric fields and radiation performances for the antennas. Some modes are inefficient for realizing wideband or multiband performance, which is needed to suppress or remove them with the excitation of these desired modes. Based on the odd- and even-periodic characteristics of the internal electric fields, these above modes could be divided in two categories: (1) odd-order modes, which keep the internal electric-field distributions with odd symmetry about the patch center; and (2) even-order modes, which keep the internal electric-field distributions with even symmetry about the patch center.

There is a significant difference in internal electric-field distribution between these two mode types. Odd-order modes, such as TM_{01}, TM_{21}, and TM_{03} in Figure 2.2(a, c, d), have common electric-field nodes at the center of the radiating patch. In contrast, the even-order TM_{02} mode exhibits an electric-field antinode at the center of the rectangular patch as illustrated in Figure 2.2(b).

To illustrate these modes more in detail, we will take the odd-order modes TM_{01} and TM_{03}, and even-order mode TM_{02} as examples. Figure 2.3 plots the schematic diagram of the input resistance responses for the conventional MPA under these three modes, and Figure 2.4 gives the schematic diagram of their internal electric-field distributions from a side view and a 3D view. For the TM_{01} mode in Figure 2.4(a), the electric-field node locates at the center of the radiating patch. For the TM_{03} mode in Figure 2.4(c), the electric-field nodes

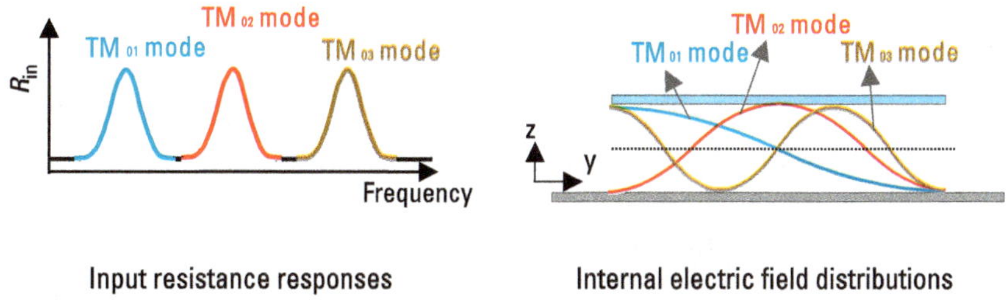

Figure 2.3 Input resistance responses and side-view electric-field distributions of the MPA under TM_{01}, TM_{02}, and TM_{03} modes.

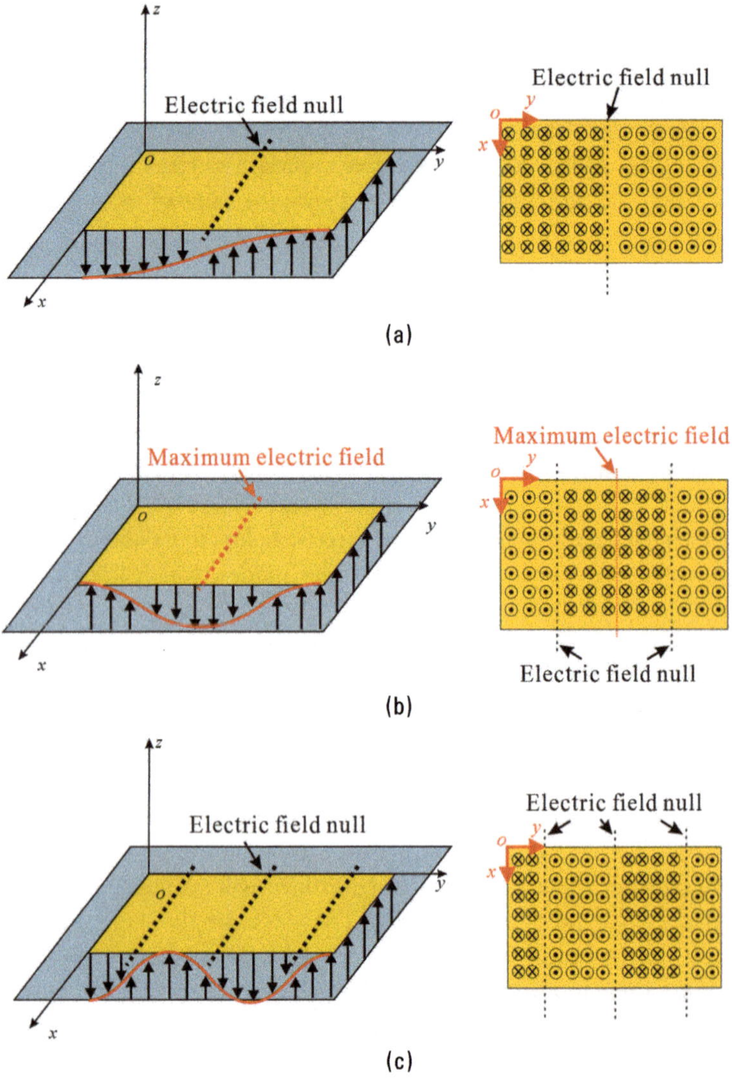

Figure 2.4 Internal electric-field distributions of the MPA in different modes: (a) TM_{01} mode, (b) TM_{02} mode, and (c) TM_{03} mode.

locate at the center, as well as at a distance with one-sixth from the edge of the radiating patch. However, for the TM_{02} mode, it can be seen from Figure 2.4(b) that its electric-field exhibits an antinode at the center of the patch.

In this view, it is possible to use the multimode differences in their electric-field distributions and introduce some special components, such as shorting pins, slots, and open stubs, to disturb the original internal electric-field distribution characteristics and ultimately achieve the effective excitation, or suppression, of these specific modes.

2.2.3 Even-Order Mode-Suppression Methods

In this part, we will introduce a working principle of how to suppress these even-order modes while keeping the excitation of these odd-order modes. In this condition, several attractive performances, such as broadside radiation beam, could be successfully gained for the MPA. According to the electric-field distribution in Figure 2.4, even-order and odd-order modes possess different symmetry about the central axis of the radiator. If we build up this position as the equivalent electric walls, all these even-order modes could be successfully suppressed for the antenna. Figure 2.5 plots the schematic diagram of the input resistance response of the antenna with excitation of these odd-order modes and suppression of these even-order modes. It also gives the schematic diagram of their corresponding internal electric-field distributions, which forms the minimum electric field at the center to retain the performance of its odd-order TM_{01} and TM_{03} modes, without the excitation of the even-order TM_{02} mode.

Figure 2.6 shows several effective methods for even-order mode suppression and odd-order mode excitation simultaneously of the MPA in the following.

First, the differential feeding scheme is an effective method to form up the equivalent electric wall at the patch center, as shown in Figure 2.6(a). A pair of feeding probes, arranged symmetrically along the y-direction, keeps the same amplitude and 180°-phase difference, counteracting the maximum internal electric-field amplitude of these even-order modes and maintaining the internal electric field of its counterpart odd-order modes.

Second, the aperture-coupled feeding is another effective method to form up the equivalent electric wall, as shown in Figure 2.6(b). A linear slot is cut on the ground plane and below the center of the radiating patch. Then a

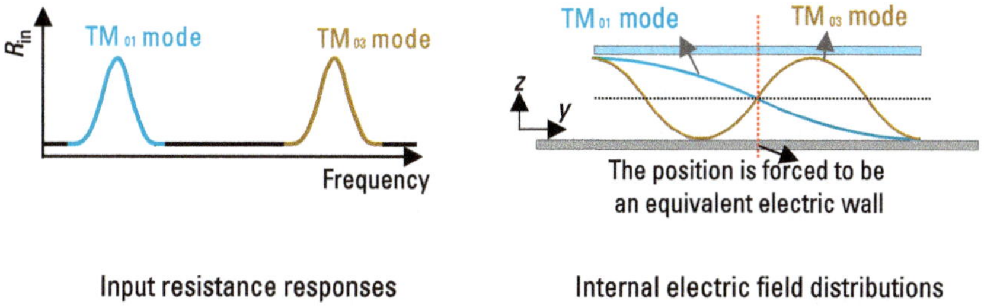

Figure 2.5 Input resistance responses and side-view electric-field distributions of a traditional MPA under even-order mode suppression.

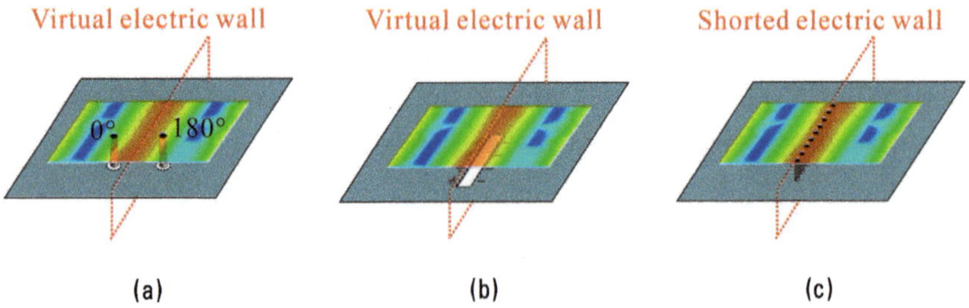

Figure 2.6 Even-order mode-suppression methods for the MPA: (a) differential feeding, (b) aperture-coupled feeding, and (c) shorting pins loading.

positive- and negative-voltage difference is formed on the sides of the aperture slot by using the microstrip feeding line below it, creating the minimum voltage at the slot center and constructing the equivalent electric wall on the radiating patch center, counteracting the maximum internal electric-field amplitude of these even-order modes and maintaining the internal electric field of its counterpart odd-order modes.

Third, shorting pins technology is an alternative method for forming up the equivalent electric wall, as shown in Figure 2.6(c). A set of shorting pins, or shorted metal, are loaded underneath the radiating patch. The internal electric fields in the patch center are directly built as the equivalent electric wall, so these even-order modes could be successfully suppressed or removed.

These above methods are only three examples for even-order mode suppression. If we try to suppress some special modes, the internal electric fields of this mode could be properly controlled via these pins, slots, stubs, and so on.

2.2.4 Odd-Order Mode-Suppression Methods

Apart from the above even-order mode suppression, MPAs with odd-order mode suppression are also valuable for special performances, such as dual radiation beams. According to the internal electric-field distributions in Figure 2.4, the difference between odd-order and even-order modes is still about the magnitude of the internal electric fields at the patch center. Figure 2.7 plots the schematic diagram of input resistance curves for the conventional MPA under the condition of odd-order TM_{01} mode and TM_{03}-mode suppression. The corresponding schematic diagram of its internal field distributions is also plotted there. If the internal electric fields at the patch center are transformed into an equivalent magnetic wall, the original internal electric-field amplitude of its even-order TM_{02} mode remains almost constant, keeping its primary resonance and radiation characteristics. However, the internal electric-field amplitudes of these odd-order TM_{01} and TM_{03} modes are destroyed, suppressing their characteristics.

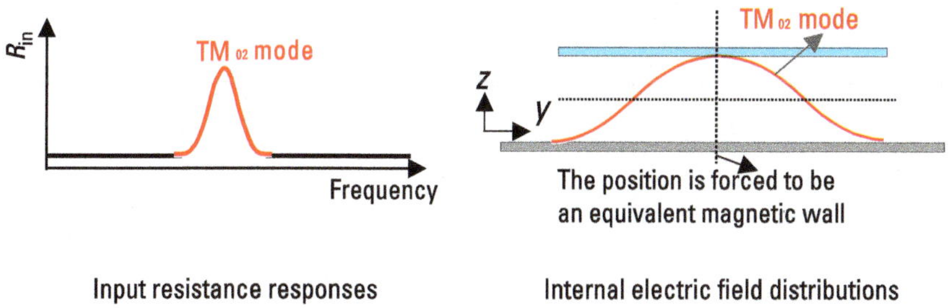

Figure 2.7 Input resistance responses and side-view electric-field distributions of a traditional MPA under odd-order mode suppression.

Figure 2.8 shows several effective methods for odd-order mode suppression and even-order mode excitation simultaneously of the MPA in the following.

First, compared with the differential feeding ports in Figure 2.8(a), we can still use two feeding ports, but they should be excited with equal amplitude and same-phase difference, as shown in Figure 2.8(a). The equivalent magnetic wall could be formed up around the center of the radiating patch, leading to the suppression of these odd-order modes.

Second, compared to the aperture-coupled feeding in Figure 2.6(b), the central feeding scheme is another effective method to form up an equivalent magnetic wall, as shown in Figure 2.8(b). The feeding port is directly loaded underneath the center of the radiation patch, resulting in the suppression of these odd-order modes and excitation of these counterpart even-order modes.

These above methods are only two examples for odd-order mode suppression. If we try to suppress some special modes, the internal electric fields of this mode could be properly controlled via these pins, slots, stubs, and so on.

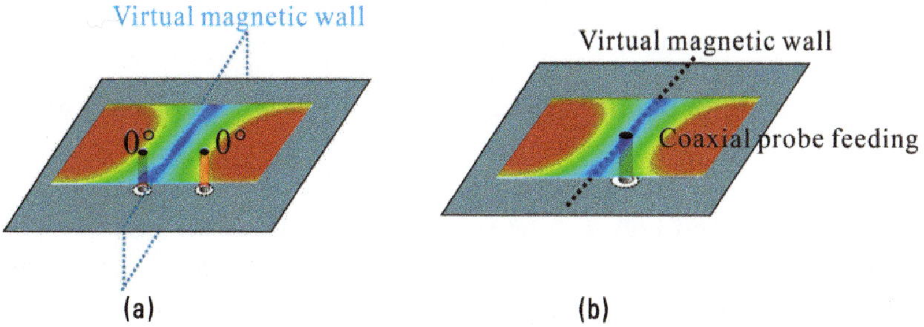

Figure 2.8 Odd-order mode-suppression methods for the MPA: (a) equal amplitude and in-phase feeding scheme, and (b) central feeding scheme.

2.3 Resonant Frequency Under Multimode Resonance

Based on (2.1), the relationship between the wave number k_{mn} and the resonant frequency f_{mn} of the MPA at TM_{mn} mode could be connected as below

$$k_{mn} = \sqrt{\left(\frac{m\pi}{a}\right)^2 + \left(\frac{n\pi}{b}\right)^2} = k_0\sqrt{\varepsilon_r} = \frac{2\pi\sqrt{\varepsilon_r}f_{mn}}{c} \tag{2.4}$$

where m and n are equal to 0, 1, 2...

According to (2.4), the resonant frequency f_{mn} of the antenna at TM_{mn} mode could be obtained as

$$f_{mn} = \frac{c}{2\sqrt{\varepsilon_r}}\sqrt{\left(\frac{m}{a}\right)^2 + \left(\frac{n}{b}\right)^2} \tag{2.5}$$

From the previous work [3, 4], the calculated resonant frequency f_{mn} from (2.5) is not accurate enough. To address this critical issue, it is necessary to use the equivalent dielectric constant ε_e instead of the relative dielectric constant ε_r. Meanwhile, the edge effect of the radiating patch should be considered by replacing the dimensions a and b as a_e and b_e, independently. Under this condition, the resonant frequency f_{mn} of the microstrip patch antenna in TM_{mn} mode should be corrected as

$$\begin{cases} f_{mn} = \dfrac{c}{2\sqrt{\varepsilon_e}}\sqrt{\left(\dfrac{m\pi}{a_e}\right)^2 + \left(\dfrac{n\pi}{b_e}\right)^2} \\ \varepsilon_e = \dfrac{1}{2}\left[\varepsilon_r + 1 + (\varepsilon_r - 1)\left(1 + \dfrac{10h}{w}\right)^{-1/2}\right] \\ a_e = a + h/2 \\ b_e = b + h/2 \end{cases} \tag{2.6}$$

where w is the width of the radiating patch on the nonresonant edges for a and b.

According to (2.6), the resonant frequencies of a traditional MPA under fundamental and higher-order modes are independently excited over the wide operating band, and all of them are proportional to the corresponding key numbers m and n. Figure 2.9 shows a large number of a traditional MPA. TM_{00} mode represents the zeroth-order mode, which can be excited by loading the shorting pins underneath the patch [5, 6]. There are multiple relationships among these radiative modes; for example, the resonant frequency of its TM_{02}

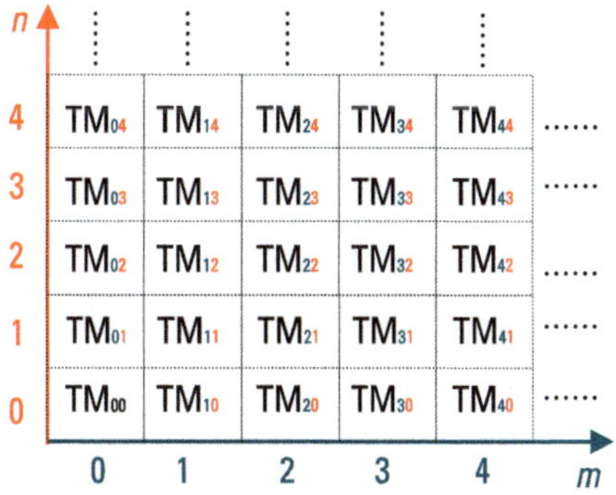

Figure 2.9 Resonant modes of a traditional MPA.

mode is twice that of the fundamental TM_{01} mode, and the resonant frequency of its TM_{40} mode is four times that of the TM_{10} mode. Tuning these modes in proximity as desired to yield multiband or wideband antennas can be difficult.

In this context, the challenging work is moved to specifically control the input resistances of the MPA, and reallocate the multiresonant modes from large frequency ratio to the desired small frequency ratio, or even merged-frequency band, achieving desired broadband, multiband, and other attractive characteristics. Researchers have carried out a series of studies on multimode planar antennas with impedance bandwidth and isolation bandwidth enhancements. Sections 2.5 and 2.6 describe the working principle of achieving these performances in greater detail.

2.4 Radiation Pattern Under Multimode Resonance

Apart from input impedances or resonant frequencies, radiation patterns are also important performance characteristics for the antenna. To show the working principle of how to control the radiation beam, Figure 2.10 illustrates the schematic diagram of the equivalent magnetic current distribution for a traditional MPA under TM_{mn} mode. Its equivalent magnetic current mainly contains $\overrightarrow{M_{s1}}$, $\overrightarrow{M_{s2}}$, $\overrightarrow{M_{s3}}$, and $\overrightarrow{M_{s4}}$, which is distributed around the edges of the radiating patch. When the length along x-direction and width along y-direction are defined as a and b, respectively, using (2.3), the equivalent magnetic current of the antenna operating in TM_{mn} mode could be expressed as

$$\overrightarrow{M_s} = -\hat{n} \times \hat{z} E_z \tag{2.7}$$

Based on (2.3) and (2.7), the $\overrightarrow{M_{s1}}$, $\overrightarrow{M_{s2}}$, $\overrightarrow{M_{s3}}$, and $\overrightarrow{M_{s4}}$, of the antenna under this boundary condition could be solved as follows

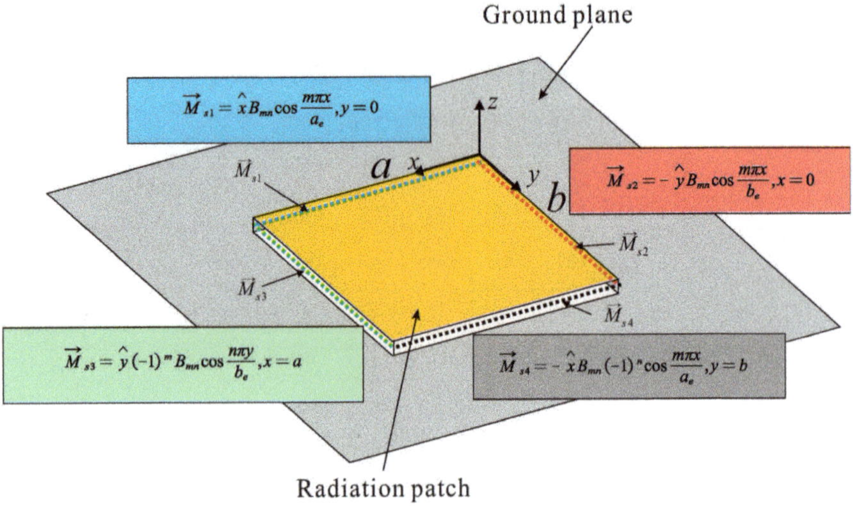

Figure 2.10 Equivalent magnetic current distributions of a traditional low-profile MPA operating in TM_{mn} mode.

$$\begin{cases} \overrightarrow{M_{s1}} = \hat{x}B_{mn}\cos\frac{m\pi x}{a_e}, y=0 \\ \overrightarrow{M_{s2}} = -\hat{y}B_{mn}\cos\frac{n\pi y}{b_e}, x=0 \\ \overrightarrow{M_{s3}} = \hat{y}B_{mn}(-1)^m\cos\frac{n\pi y}{b_e}, x=a \\ \overrightarrow{M_{s4}} = -\hat{x}B_{mn}(-1)^n\cos\frac{m\pi x}{a_e}, y=b \end{cases} \tag{2.8}$$

Therefore, the vector $\vec{F}$ from these equivalent magnetic currents $\overrightarrow{M_{s1}}$, $\overrightarrow{M_{s2}}$, $\overrightarrow{M_{s3}}$, and $\overrightarrow{M_{s4}}$ can be calculated as

$$\vec{F} = \frac{h}{2\pi r}e^{-jk_0 r}\int_s \overrightarrow{M_s}\, e^{jk_0(x\sin\theta\cos\varphi + y\sin\theta\sin\varphi)}dxdy \tag{2.9}$$

According to the vector $\vec{F}$, the far-zone-radiated fields of the antenna could be solved as

$$\vec{E} = -\nabla\times\vec{F} = jk(\hat{\varphi}F_\theta - \hat{\theta}F_\varphi) \tag{2.10}$$

$$\vec{E} = \hat{\theta}E_\theta + \hat{\varphi}E_\varphi \tag{2.11}$$

Based on (2.11), the far-zone field components E_θ and E_ϕ are calculated as

$$\begin{cases} E_\theta = jk_0(F_x \sin\varphi - F_y \cos\varphi) \\ E_\varphi = jk_0(F_x \cos\varphi + F_y \sin\varphi)\cos\theta \end{cases} \tag{2.12}$$

To compare the beam difference between these modes, the far-zone-radiated fields of the antenna at TM_{01}, TM_{02}, TM_{21}, and TM_{03} modes are calculated and normalized in the following

$$TM_{01}\begin{cases} xoz: \quad E_\theta = 0, \ E_\varphi = \cos\theta\left(\dfrac{\sin(X)}{X}\right) \\ yoz: \quad E_\theta = \cos(0.5k_0b_e\sin\theta), \ E_\varphi = 0 \end{cases} \tag{2.13}$$

$$TM_{02}\begin{cases} xoz: \quad E_\theta = 0, \quad E_\varphi = 0 \\ yoz: \quad E_\theta = \sin(0.5k_0b_e\sin\theta), \ E_\varphi = 0 \end{cases} \tag{2.14}$$

$$TM_{21}\begin{cases} xoz: \quad E_\theta = 0, \ E_\varphi = \cos(\theta)\dfrac{X\sin(X)a_e}{\pi^2 - X^2} \\ yoz: \quad E_\theta = 0, \quad E_\varphi = 0 \end{cases} \tag{2.15}$$

$$TM_{03}\begin{cases} xoz: \quad E_\theta = 0, \ E_\varphi = \cos\theta\left(\dfrac{\sin(X)}{X}\right) \\ yoz: \quad E_\theta = \cos(0.5k_0b_e\sin\theta), \ E_\varphi = 0 \end{cases} \tag{2.16}$$

where X in (2.13) to (2.16) is

$$X = \frac{a_e\pi\sin\theta}{\lambda} \tag{2.17}$$

In (2.13) to (2.17), a_e represents the length of the radiating patch with consideration of edge effects, and b_e represents the width of the radiating patch with consideration of edge effects. The corresponding equivalent magnetic current distributions and E-plane-normalized electric-field components of the antenna are plotted in Figure 2.11 under operation of TM_{01}, TM_{02}, TM_{21}, and TM_{03} modes.

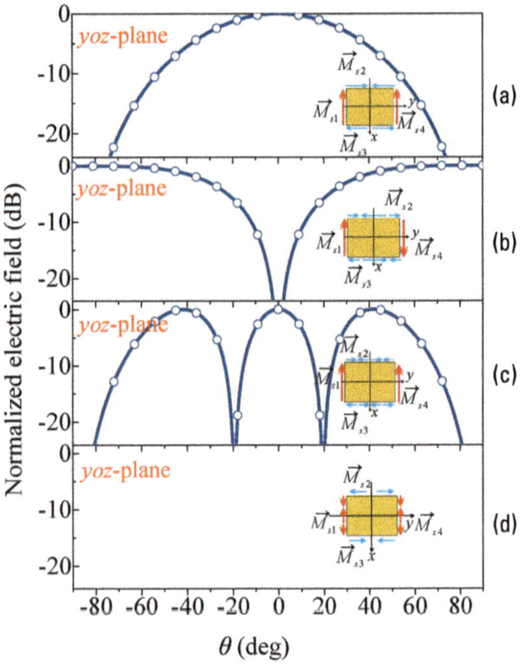

Figure 2.11 Equivalent magnetic current distribution and *E*-plane-normalized electric-field components of the antenna in Figure 2.10 under radiation of four different modes: (a) TM_{01} mode, (b) TM_{02} mode, (c) TM_{03} mode, and (d) TM_{21} mode.

Figure 2.11 shows the equivalent magnetic currents and radiation pattern characteristics of the antenna under different modes. We would like to illustrate these performances in more detail.

First, when the MPA is operated at TM_{01} mode, its *E*-plane radiation pattern is mainly generated by the red equivalent magnetic currents $\overrightarrow{M_{s1}}$ and $\overrightarrow{M_{s4}}$ along the *x*-direction with equal amplitude and same direction. Due to the small spacing between these two equivalent magnetic currents, its *E*-plane radiation pattern maintains the broadside beam without sidelobe level. At the same time, the blue equivalent magnetic currents $\overrightarrow{M_{s2}}$ and $\overrightarrow{M_{s3}}$ arranged along the *y*-direction keep the opposite direction, and cancel each other, so that there are no radiated fields from them, as shown in Figure 2.11(a). If a single feeding port is introduced on one side of the radiating patch, these blue equivalent magnetic currents will keep imbalance performance so as to introduce the cross polarization in E-plane.

Second, when the MPA is operated at TM_{02} mode, its *E*-plane radiation pattern is still mainly generated by the red equivalent magnetic currents $\overrightarrow{M_{s1}}$ and $\overrightarrow{M_{s4}}$ along the *x*-direction. However, due to the opposite direction between them, the *E*-plane radiated fields produce the null in the broadside direction.

Meanwhile, the blue equivalent magnetic currents $\overrightarrow{M_{s2}}$ and $\overrightarrow{M_{s3}}$, arranged along the *y*-direction, are also opposite and cancel each other, so that these equivalent magnetic currents also generate no radiated fields at the *E*-plane, as shown in Figure 2.11(b). If a single feeding port is introduced on one side of the radiating patch, these blue equivalent magnetic currents will keep imbalance performance, leading to an increased cross-polarization level at *E*-plane.

Third, when the MPA is operated at TM_{03} mode, its *E*-plane radiation pattern is also generated by the red equivalent magnetic currents $\overrightarrow{M_{s1}}$ and $\overrightarrow{M_{s4}}$ along the *x*-direction with equal amplitude and in-phase difference. But the distance between these parallel equivalent magnetic currents of the TM_{03} mode is about three times larger than that of TM_{01} mode, resulting in the high sidelobe levels at *E*-plane pattern as shown in Figure 2.11(c). With respect to its blue equivalent magnetic currents $\overrightarrow{M_{s2}}$ and $\overrightarrow{M_{s3}}$ along the *y*-direction, a similar performance is produced for the antenna, and if a single feeding port is introduced on one side of the radiating patch, these blue equivalent magnetic currents will keep imbalance performance, leading to an increased cross-polarization level in this plane.

Fourth, when the antenna is operated at TM_{21} mode, the red equivalent magnetic currents $\overrightarrow{M_{s1}}$ and $\overrightarrow{M_{s4}}$ arranged along the *x*-axis are cancelled out, resulting in no radiation fields at *E*-plane. At the same time, the blue equivalent magnetic currents $\overrightarrow{M_{s2}}$ and $\overrightarrow{M_{s3}}$ arranged along the *x*-direction are opposite and cancel each other, so there are no radiated fields at *E*-plane radiation pattern, as shown in Figure 2.11(d). Researchers have often ignored this mode and have defined it as a nonuseful pattern. However, if we properly adjust the magnitude of these equivalent magnetic current distributions, the nonbroadside radiation pattern could be transformed into the broadside beam, and finally, its reshaped *E*-plane radiation pattern could be similarly shaped with that of its TM_{01} mode, and ultimately could be turned into the useful mode for wide bandwidth or multibandwidth performances.

To display the radiation patterns of the antenna at TM_{01}, TM_{02}, TM_{21}, and TM_{03} modes more clearly, Figure 2.12 shows the 3D patterns of the antenna at these modes. When the antenna is operated at the fundamental TM_{01} mode, it has the desired directional radiation pattern without sidelobe, as shown in Figure 2.12(a). This mode is commonly selected and investigated in the design of high-performance antennas. When the antenna is operated at higher-order TM_{02} mode, it suffers from the null in the broadside direction, as shown in Figure 2.12(b). When the antenna is operated at higher-order TM_{03} mode, it has a critical issue of high sidelobe level due to the large spacing at E-plane, as shown in Figure 2.12(c). When the antenna is operated at high-order TM_{21} mode, the peak gain is maintained off-broadside direction, as shown in Figure 2.12(d).

Another challenge for mode combination is the significant difference in radiation patterns. To make full use of multimode characteristics, it is necessary

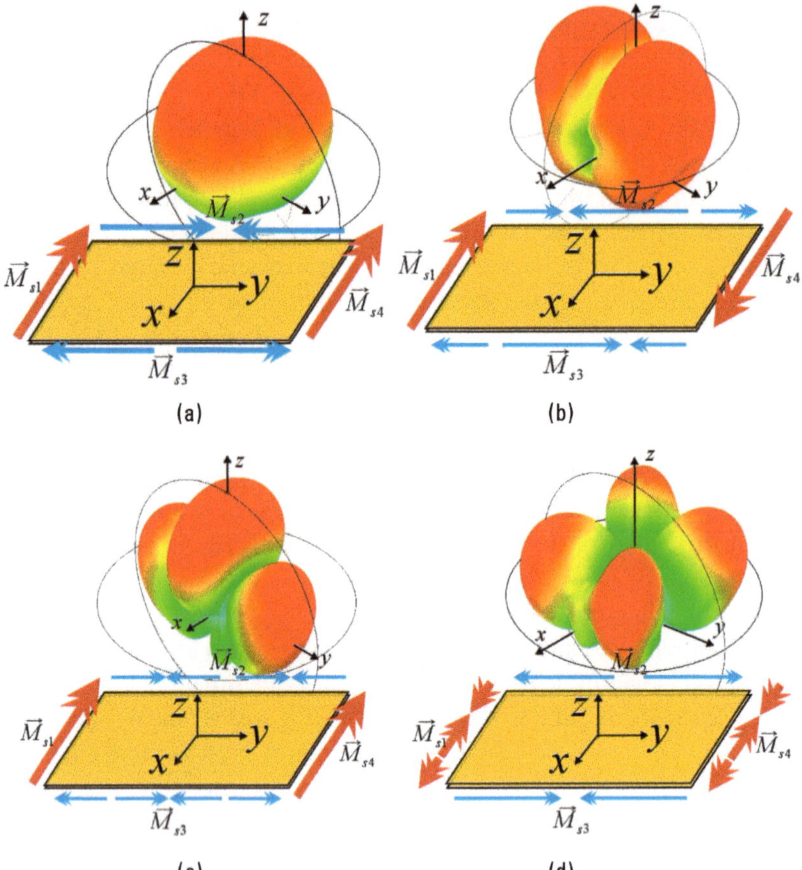

Figure 2.12 3D-view radiation patterns of a traditional MPA at different modes: (a) TM_{01} mode, (b) TM_{02} mode, (c) TM_{03} mode, and (d) TM_{21} mode.

to consider the radiation pattern control of these higher-order modes. How to achieve this performance is described in more detail in Sections 2.7 and 2.8.

2.5 Impedance Bandwidth Improvement

Figure 2.9 shows a large number of radiative modes excited over the wide operating band. To realize wide bandwidth or multibandwidth by making use of these modes, we must control the resonant frequencies of these modes independently. Figure 2.13 plots the schematic diagram of the input resistances at mode 1 and mode 2. Initially, for a traditional antenna, the resonant frequency f_1 of its mode 1 keeps the large frequency ratio with that of mode 2 (f_2), as shown before controlling. To realize the wide bandwidth, the resonant frequency f_1 of the

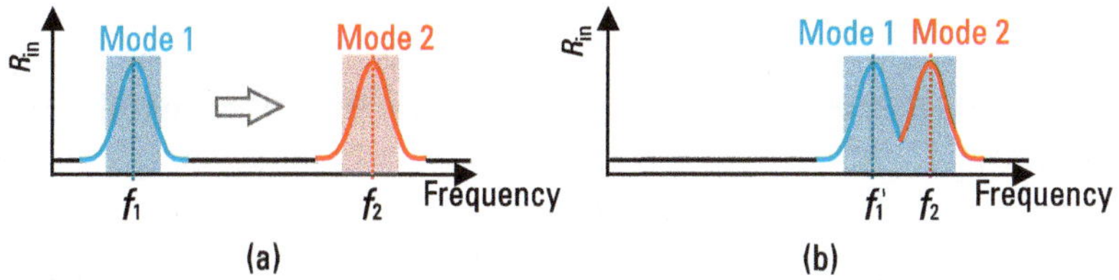

Figure 2.13 Input resistance controlling of a traditional MPA: (a) before controlling, and (b) after controlling.

antenna at mode 1 should be effectively pushed up to about f_1' while maintaining a stable resonant frequency f_2 at its mode 2, forming a dual-minimum-pole characteristic and effectively widening the operating bandwidth of the antenna, as shown after controlling. This method can also be used for the design and implementation of multiband antenna with the controllable frequency ratio.

Next, the resonant frequency of these modes could be shifted by disturbing the internal electric-field distributions. The equivalent circuit of a traditional MPA loaded with the inductors L_s is established in Figure 2.14. Herein, the TM_{01} and TM_{02} modes are selected as an example to analyze the resonant frequencies of these modes. From this figure, the inductor is loaded at the zero-voltage position of its TM_{02} mode (the position where the magnitude of its internal electric field keeps the minimum value), but this place is about the relative maximum internal electric fields of its TM_{01} mode, which keeps far away from the zero-voltage position. Under this condition, the resonant frequency of the antenna at fundamental TM_{01} mode is effectively increased up, while the

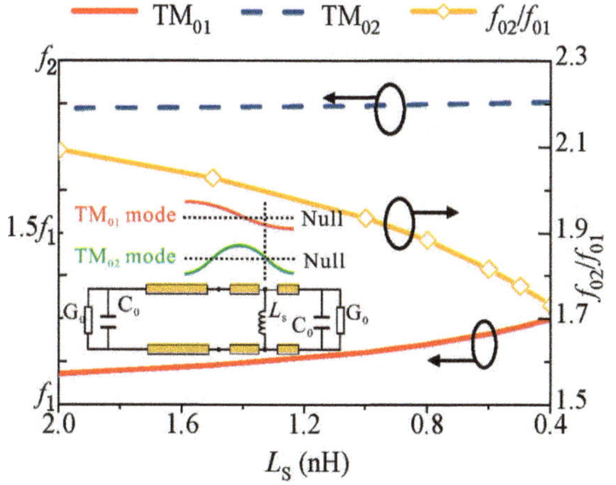

Figure 2.14 Schematic diagram of equivalent transmission-line model for a traditional MPA based on the loaded inductance L_s.

resonant frequency of the antenna at higher-order TM_{02} mode remains almost stable. As such, the frequency ratio between these dual modes could be effectively controlled to the desired value for wide bandwidth or multibandwidth performances.

To be extended, Figure 2.15 plots the schematic diagram of the input resistances at mode 1 and mode 2. For a traditional antenna, the resonance ratio between its mode 1 and mode 2 are maintained far away from each other, as shown before controlling. Then, in order to realize the wide bandwidth with compact size, the resonant frequency f_2 of the antenna at mode 2 could be effectively moved down to about f_2', while maintaining a basically stable resonant frequency f_1 at its mode 1, as shown after controlling.

Based on the scheme in Figure 2.15, the equivalent circuit of a traditional MPA loaded with the capacitance C_s is established in Figure 2.16. The TM_{01} and TM_{02} modes are still selected as an example to analyze the resonant frequencies of these modes. For the TM_{02} mode, the capacitor is loaded at its zero-voltage position (where the magnitude of its internal electric field keeps the minimum value), but for the TM_{01} mode, the capacitor is loaded about the relative maximum internal electric fields, which keeps far away from the zero-voltage position. Under this condition, the resonant frequency of the TM_{02} mode is moved down, while the resonant frequency of the TM_{01} mode remains almost constant. The frequency ratio between these dual modes could be effectively controlled and reduced to the desired value for wide bandwidth or multibandwidth performance.

2.6 Isolation Bandwidth Improvement

Apart from the impedance bandwidth, the isolation bandwidth of the antenna is also important, especially in multiple-input multiple-output (MIMO) communication systems. An isolation improvement based on the multimode concept has been investigated and developed thoroughly in recent years, especially in a single-sharing radiator.

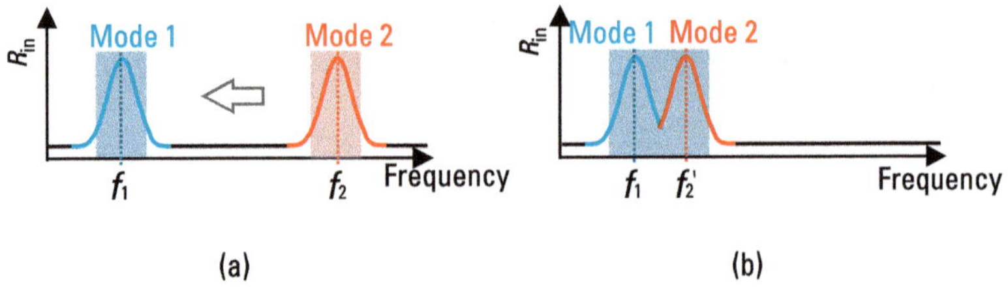

Figure 2.15 Input resistance controlling of a traditional MPA: (a) before controlling, and (b) after controlling.

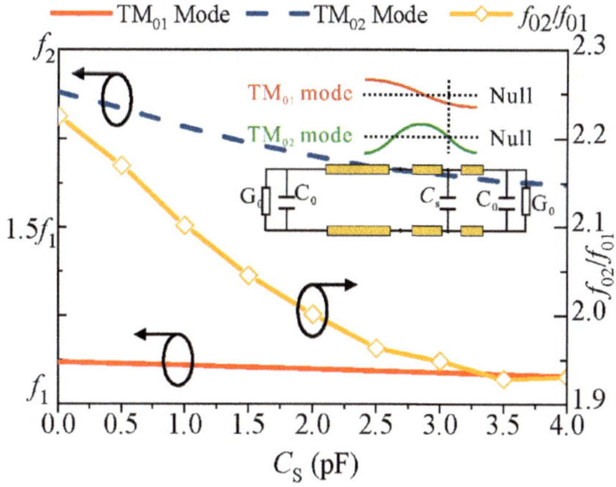

Figure 2.16 Schematic diagram of equivalent transmission-line model for a traditional MPA based on the loaded capacitance C_S.

Figure 2.17 plots the configuration of a traditional single MPA under dual ports; a set of modes could be excited over the wide operating band. Here, we take mode 1 and mode 2 as an example for working principle discussion. For a traditional multiport antenna with a single mode, although the input impedance of these dual modes is kept as small (below –10 dB) as desired, the antenna suffers from poor isolation. To address this issue, the resonant frequencies of mode 1 and mode 2 are reallocated and moved in proximity to each other. A minimum pole for the S_{12} is generated for the antenna. If we combine more modes, the minimum poles of S_{12} will be increased further, achieving an improvement in isolation bandwidth between these multiports.

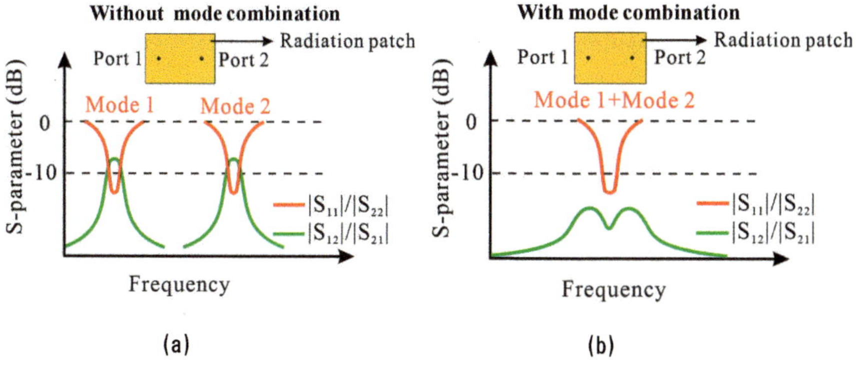

Figure 2.17 Schematic diagram of isolation improvement based on two modes of combination: (a) two separate modes for a traditional MPA, and (b) two combined modes for the improved MPA.

Next, to illustrate this working principle in more detail, we need to use some special modes for isolation improvement. We have selected the combination of TM_{01} and TM_{02} modes, and the combination of TM_{03} and TM_{04} modes. The odd-order mode and even-order mode should be selected together, the basic requirement for multimode-isolation improvement. Based on the internal electric-field distributions of the MPA in Section 2.2, Figure 2.18 gives the electric-field distribution of the antenna at TM_{01} mode under different feeding port 1. When the feeding port 1 is excited on the left side of the radiating patch, the electric field on the left side of the patch is directed downward, while the electric field on the right side of the patch is directed upward. The dashed line in the center represents the node of its electric fields, which keep an almost constant value along the x-axis. By symmetry, as the counterpart port 2 is excited, the electric-field distributions on the right side can be given. In this case, an upward electric field is formed on the left side of the patch, and a downward electric field is formed on the right side of the patch. The electric-field distribution between these two port excitations independently keeps the opposite direction.

Similar to the electric-field distribution of the TM_{01} mode, the electric-field distributions of its TM_{02} mode under excitation of port 1 and port 2 are also depicted in Figure 2.18. As the left port 1 is excited, the downward electric-field distribution is formed in the left quarter and right quarter of the radiating patch, and an upward electric field is generated at the middle half of the radiating patch. As the right port 2 is excited, the electric-field distribution on the radiating patch can be directly obtained based on the symmetrical scheme. The electric-field distribution between these two port excitations keeps the same direction.

Based on the above electric-field distributions for its TM_{01} and TM_{02} modes, the combined electric fields from these two modes are further discussed in the following. Figure 2.18 shows the relevant results of the combined modes under different magnitude ratios with the excitation of port 1 or port 2 independently, where the magnitude ratios between TM_{01} mode and TM_{02} mode are set as 1:4, 1:2, and 1:1 under the excited port 1, respectively. Initially, when the magnitude ratio is selected as 1:4, the electric-field contribution from the TM_{01} mode is relatively small, and the electric-field contribution is mainly determined by the TM_{02} mode. The electric-field node is distributed around the right quarter of the radiating patch. By increasing the magnitude ratio, the electric-field contribution from the TM_{01} mode is gradually increased, so its electric-field node is progressively moved to the right side. As the magnitude ratio reaches to about 1:1, the electric-field node will be reallocated in proximity to the right-side edge. Due to the symmetrical structure, the electric-field node of the combined two modes is progressively moved to the left-side edge under the excitation of port 2 independently.

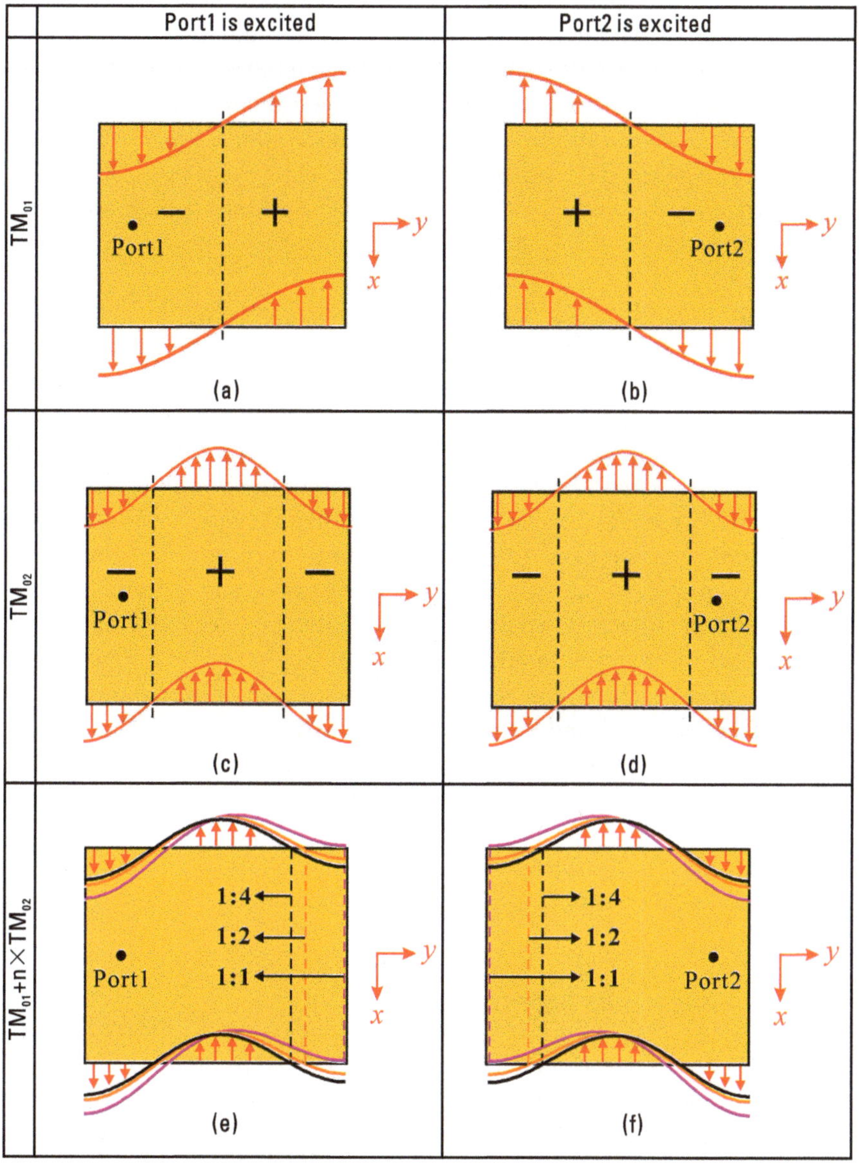

Figure 2.18 Electric-field distribution of the MPA at TM_{01} and TM_{02} modes under different excitation of the feeding port: (a) TM_{01} mode under excitation of port 1, (b) TM_{01} mode under excitation of port 2, (c) TM_{02} mode under excitation of port 1, (d) TM_{02} mode under excitation of port 2, (e) TM_{01} and TM_{02} mode combination under excitation of port 1, and (f) TM_{01} and TM_{02} mode combination under excitation of port 2.

If we properly select the unexcited port around the position of the electric-field nodes, the minimum pole for the S_{12} is generated during the working band.

Similar to the above principle, the electric-field distribution of the higher-order TM_{03} and TM_{04} modes can be conducted and discussed, with relevant results given in Figure 2.19. This figure shows that the electric fields of its TM_{03} mode produce the positive- and negative-voltage variation about three times along the y-axis, while the electric field of its TM_{04} mode keeps the positive- and negative-voltage variation about four times along the y-axis. The electric-field distribution between these two port excitations keeps the opposite direction under odd-order TM_{03} mode, but the same direction under even-order TM_{04} mode.

At this point it is necessary for us to combine the electric-field distribution of TM_{03} and TM_{04} modes in proximity to each other. Figure 2.20 plots the electric-field node variation of the combined modes under port 1 and port 2 independently. As port 1 is excited, the electric-field node is gradually moved to the right side by reducing the magnitude between these modes. In contrast, when port 2 is excited, the electric-field node from the combined two modes is progressively moved to the left side by decreasing the magnitude ratio. If we properly select the unexcited port around the position of the electric-field node, the minimum pole for the S_{12} is generated during the working band.

Due to the similar electric-field node variation between these four modes, we can apply this method to other higher-order modes, and combine them in proximity to each other. With these arrangements, a set of minimum poles for S_{12} could be generated under this condition. Figure 2.20 plots the schematic diagram of these four modes combination, where the TM_{01}, TM_{02}, TM_{03}, and TM_{04} modes are presented as the commonly used modes with both even-order and odd-order electric-field performances. These four modes are kept far away from each other. By combining them in proximity, the isolation bandwidth of the antenna could be enhanced as compared to the results in Figure 2.17.

2.7 Gain Bandwidth Improvement

Figure 2.11 shows that the antenna at TM_{01} mode keeps the HPBW of about 60° to 90° due to the key factor b_e; and the radiation pattern of its higher-order TM_{03} mode suffers from the high sidelobe level due to the large factor b_e, which reduces the radiation gain in the broadside direction. This requires us to enhance the gain of the antenna. The array technique based on multielement is the most commonly used method for gain enhancement, but the gain of the antenna element is slightly increased. This part will take the TM_{01} mode of the antenna as an example to illustrate the gain enhancement or HPBW enhancement.

Figure 2.21 shows the HPBW variation of the MPA under TM_{01} mode at different b_e. This figure shows that the electrical length b_e of a traditional

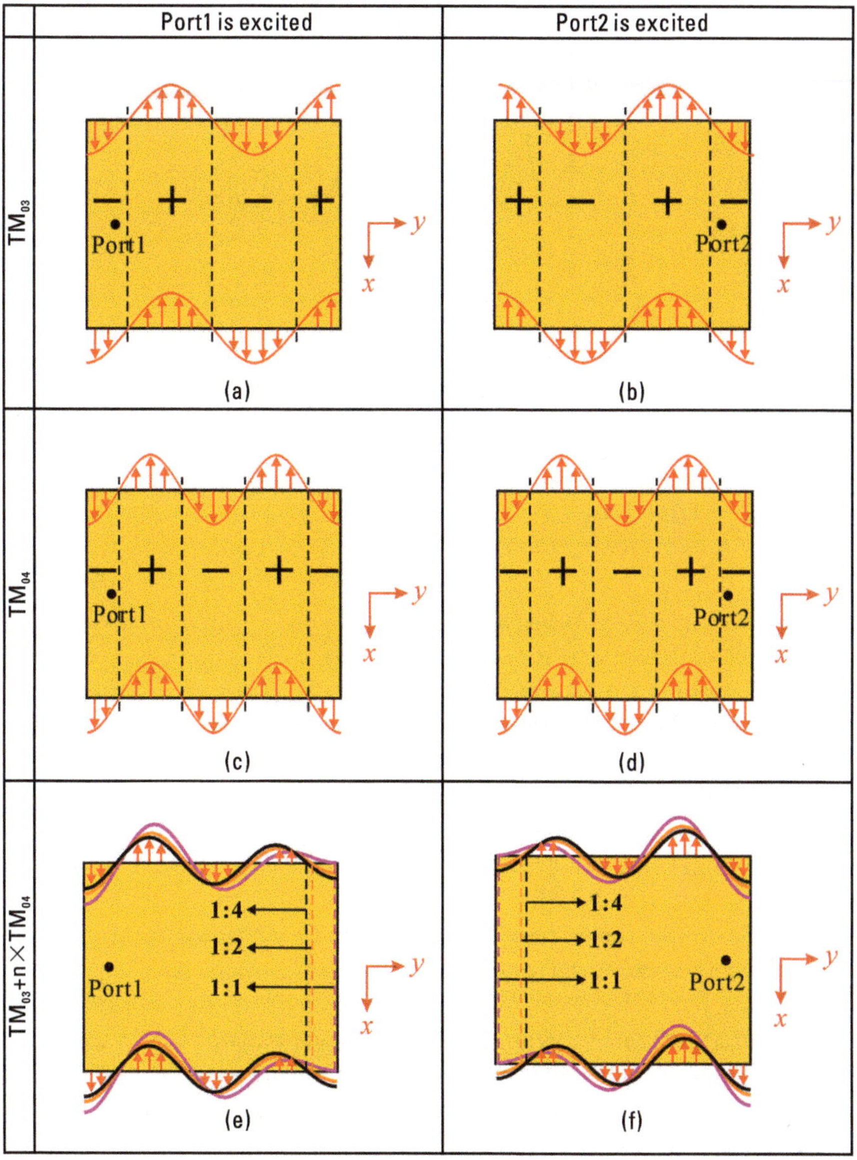

Figure 2.19 Electric-field distribution of the MPA at TM_{03} mode and under TM_{04} mode different excitation of the feeding port: (a) TM_{03} mode under excitation of port 1, (b) TM_{03} mode under excitation of port 2, (c) TM_{04} mode under excitation of port 1, (d) TM_{04} mode under excitation of port 2, (e) TM_{03} and TM_{04} mode combination under excitation of port 1, and (f) TM_{03} and TM_{04} mode combination under excitation of port 2.

MPA is usually kept around $0.5\lambda_0$, resulting in the HPBW of around 60°, as shown in Figure 2.21(a). Next, with the increasing value of the electrical length b_e, the HPBW of the antenna is gradually decreased and the sidelobe level is

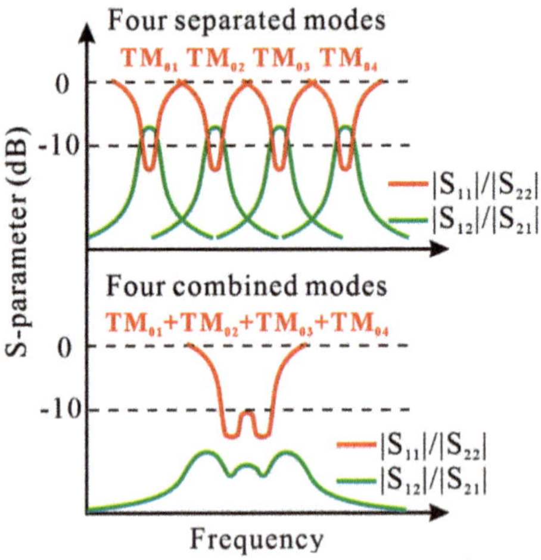

Figure 2.20 Schematic diagram of isolation improvement between four modes combination: four separated modes and four combined modes.

progressively increased. When the electrical length b_e reaches to about $0.6\lambda_0$, the HPBW of the antenna is compressed to about 49° with the acceptable sidelobe characteristics, as shown in Figure 2.21(b, c). When the electrical length b_e increases to about $1.0\lambda_0$, the HPBW of the antenna is further compressed to about 29°, but the sidelobe level is increased and becomes similar to the broadside beam level, as shown in Figure 2.21(d).

If we try to realize the increased gain for the proposed MPA, the electrical length b_e should be controlled and kept at around $0.6\lambda_0$ as desired.

Apart from the above method, the additional radiation element could be introduced around the primary element for HPBW enhancement. Figure 2.22 gives the comparison diagram of the MPA at TM_{01} mode with and without loading the linear slot. From the figure, before loading the slot, the *E*-plane HPBW of a traditional antenna is relatively large, but when the linear slot is introduced on the radiating patch, the antenna not only generates the copolarized radiated fields from the equivalent magnetic currents $\overrightarrow{M_{s1}}$ and $\overrightarrow{M_{s2}}$, but generates the copolarized radiated fields through the equivalent magnetic current $\overrightarrow{M_{s3}}$ from the linear slot. Most important, the $\overrightarrow{M_{s3}}$ and $\overrightarrow{M_{s1}}$ (or $\overrightarrow{M_{s2}}$) keeps the opposite direction. Under this condition, the *E*-plane HPBW of the antenna is compressed with low sidelobe level characteristics, achieving high gain in the broadside direction as desired.

Except for the fundamental TM_{01} mode, the employment of higher-order modes is an alternative method for gain enhancement, since they have the relative narrow HPBW with a set of high sidelobe level. If we could reduce these sidelobe levels, the gain of the antenna would be further improved. In sum-

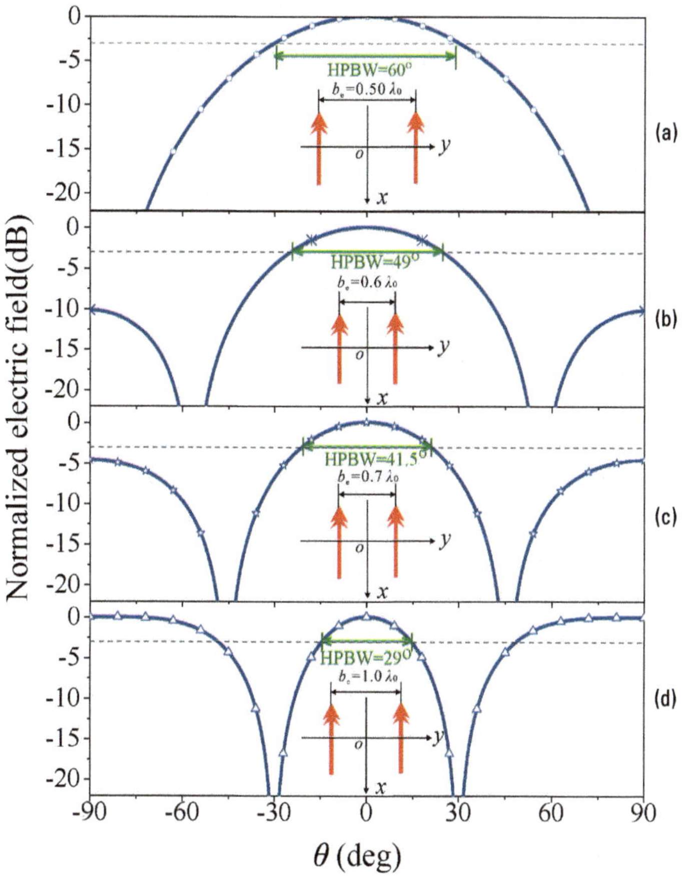

Figure 2.21 E-plane HPBW variation of the MPA at TM_{01} mode under four different values of b_e: (a) $b_e = 0.5\lambda_0$, (b) $b_e = 0.6\lambda_0$, (c) $b_e = 0.7\lambda_0$, and (d) $b_e = 1.0\lambda_0$.

mary, to acquire high-gain characteristics, some key parameters, such as array spacing b_e, affect the HPBW of the antenna and should be controlled properly.

2.8 Pattern Bandwidth Improvement

An antenna with wide HPBW coverage is also valuable in wireless communications. Here, the TM_{01} mode of the MPA is still taken as an example for HPBW discussion. According to (2.18), the E-plane radiated field E_θ and E_ϕ components of the antenna in TM_{01} mode can be expressed as

$$E_\theta = \cos(0.5 k_0 b_e \sin\theta), \quad E_\varphi = 0 \tag{2.18}$$

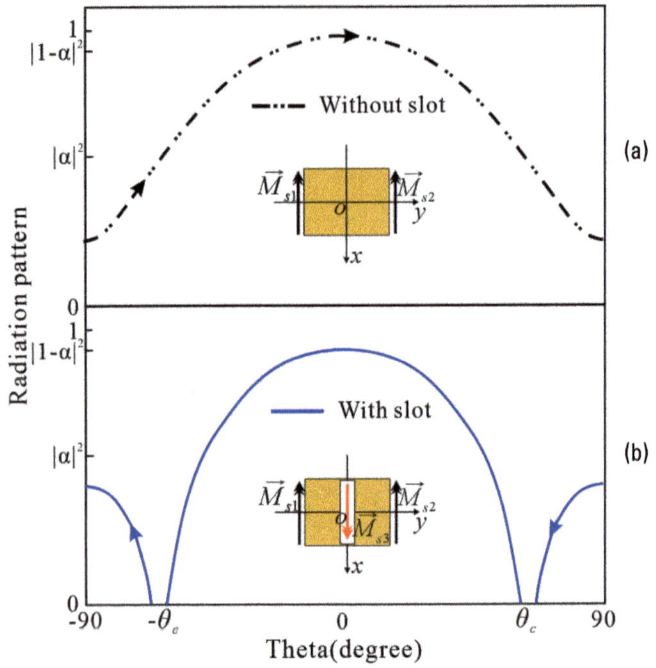

Figure 2.22 E-plane radiation pattern comparison of the traditional MPA at TM_{01} mode with and without the loaded linear slot at the patch center: (a) without the slot, and (b) with the slot.

Based on (2.18), the *E*-plane HPBW of the antenna at TM_{01} mode is also varied with the spacing b_e and wavelength.

Based on (2.18), Figure 2.23 shows the E-plane HPBW variation of the antenna under operation of TM_{01} mode as a function of different b_e. This figure shows that the electrical length b_e for a traditional MPA is usually maintained around $0.35\lambda_0$ to $0.5\lambda_0$, resulting in the HPBW of around 60° to 90° as mentioned above; the relevant results are shown in Figure 2.23(a, b). By reducing the electrical length b_e, the HPBW of the antenna could be gradually increased. When the electrical length b_e is moved to about $0.25\lambda_0$, the HPBW of the antenna is enlarged to about 173°, as shown in Figure 2.23(c). When the electrical length b_e is reduced to about 0, these two red equivalent magnetic currents are completely overlapped and replaced by a single one for far-zone radiated field generation. At this time, the HPBW of the antenna is close to 180°, as shown in Figure 2.23(d).

Compared to increasing the E-plane HPBW, increasing the H-plane HPBW of a planar antenna is more challenging. First, for the fundamental TM_{01} mode or higher-order TM_{03} mode, the equivalent magnetic currents at H-plane are only considered as a single one, so the H-plane HPBW reduction of the antenna is only realized by shortening the length of the equivalent

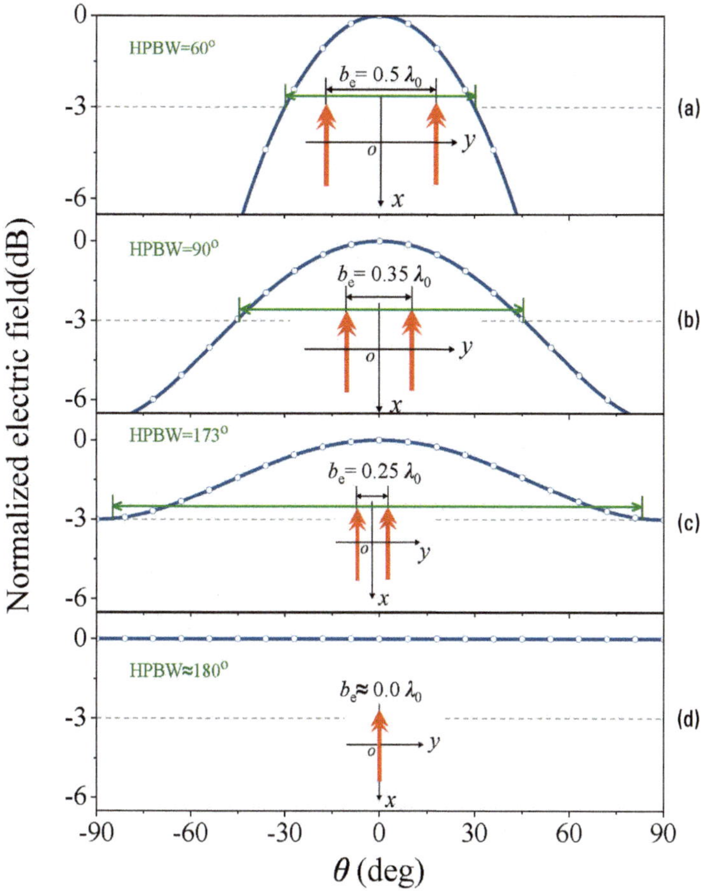

Figure 2.23 Radiation pattern variation of the MPA at TM_{01} mode under four different values of b_e: (a) $b_e = 0.5\lambda_0$, (b) $b_e = 0.35\lambda_0$, (c) $b_e = 0.25\lambda_0$, and (d) $b_e \approx 0.0\lambda_0$.

magnetic currents, which is about the width of the radiating patch. Second, the ground plane could be folded, enhancing the gain off-broadside direction and widening the HPBW of the antenna at the H-plane radiation pattern [7]. Third, vertically oriented currents could be introduced around the edges of the radiating patch to increase the gain off-broadside direction [8]. Fourth, the magnetoelectric (ME) dipole is believed to be an effective approach for both E- and H-plane HPBW enhancement, which could increase these values to about 120° [9]. However, some of these methods will increase antenna thickness. Several other effective methods, not illustrated in detail in this chapter, have been proposed and developed during the past two decades.

The cross-polarization level of an antenna is also important, so we will introduce a working principle for reducing cross-polarization levels.

With respect to a traditional MPA, the calculated results in Figure 2.11 show that there are no cross-polarized components for the antenna due to the ideal and balance equivalent magnetic currents. In fact, the asymmetric feeding port for the fabricated MPA leads to an imbalance for the equivalent magnetic currents, increasing the cross-polarization level, especially at H-plane. To address this issue, the differential feeding scheme or aperture-coupled feeding scheme are introduced to form up the symmetric equivalent magnetic current, to significantly reduce the cross-polarization level.

The shorted patch antenna, which is about half the length of a traditional MPA, suffers from more serious cross-polarization level at H-plane due to the asymmetric structure or equivalent magnetic currents. To address this issue, the design method of the antenna is given in Figure 2.24. For a traditional shorted MPA, the red equivalent magnetic currents along x-direction keep the opposite direction. But they are asymmetric at H-plane so as to produce the high cross-polarization level of about –2.9 dB at H-plane. For the improved shorted MPA, the magnitude of the red equivalent magnetic currents along x-direction is reduced to achieve the low cross-polarization level of less than –18 dB at H-plane.

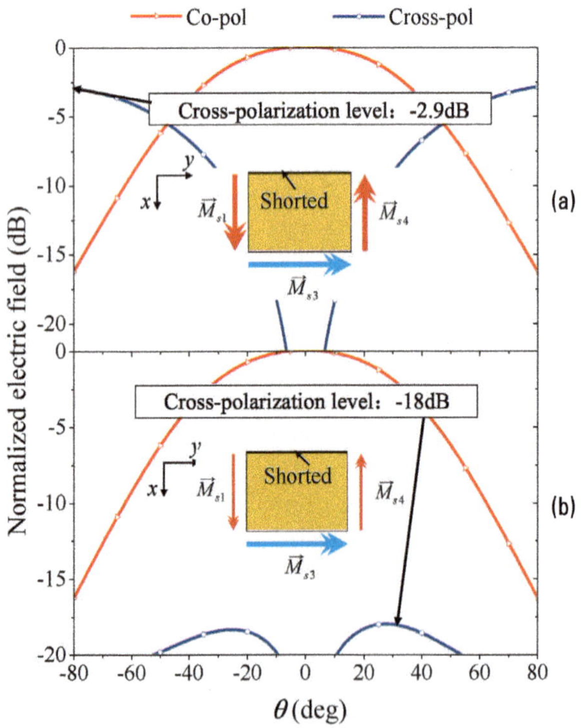

Figure 2.24 Cross-polarization reduction of the shorted MPA under different conditions: (a) without cross-polarization reduction, and (b) with cross-polarization reduction.

As reported in [10, 11], the magnitude of the red equivalent magnetic currents could be controlled by loading the shorting metal and open-ended stubs. Simulated and measured results have shown that the H-plane cross-polarization reduction reaches about −13 dB compared to a traditional counterpart.

To increase the HPBW of the antenna at E-plane, the key parameter is the array factor b_e between these parallel equivalent magnetic currents, which could realize the desired wide E-plane HPBW of about 180°. Meanwhile, to increase the HPBW of the antenna at H-plane, the additional resonant modes or vertically oriented field components should be introduced, but the H-plane HPBW of the antenna only reaches to about 120° as desired. It is still more challenging for us to design the antenna with wide H-plane HPBW, especially in a low-profile configuration. Additionally, in order to reduce the cross-polarization level, the equivalent magnetic currents of the radiative modes should be properly controlled. Some other radiation-pattern requirements, such as null scanning, filtering gain, pattern diversity, and so on, are valuable in modern wireless communications.

2.9 Concluding Remarks

This chapter introduced the mode-inspired theory of multimode planar antennas. Here, the microstrip patch antenna is taken as an example for theoretical discussion, such as internal electric-field distributions, even-order mode and odd-order mode characteristics, even-order mode suppression, odd-order mode suppression, resonant frequency of these modes, radiation pattern of these modes, impedance bandwidth enhancement, isolation bandwidth enhancement, gain bandwidth enhancement, pattern bandwidth enhancement, and so on. The above discussions provide theoretical and methodological guidance for the design of high-performance antennas in Chapters 3 through 6.

References

[1] Lo, Y., D. Solomon, and W. Richards, "Theory and Experiment on Microstrip Antennas," *IEEE Transactions on Antennas and Propagation,* Vol. 27, No. 2, March 1979, pp. 137–145.

[2] James, J. R., and P. S. Hall, *Handbook of Microstrip Antennas*, London, UK: Peregrinus, 1989.

[3] Garg, R., et al., *Microstrip Antenna Design Handbook*, Norwood, MA: Artech House, 2001.

[4] Wang, C. -F., and Y. Chen, *Characteristic Modes: Theory and Applications in Antenna Engineering*, Hoboken, NJ: Wiley, 2015.

[5] Liu, Z. -X., L. Zhu, and N. -W. Liu, "Design Approach for Compact Dual-Band Dual-Mode Patch Antenna with Flexible Frequency Ratio," *IEEE Transactions on Antennas and Propagation,* Vol. 68, No. 8, August 2020, pp. 6401–6406.

[6] Lai, A., K. M. K. H. Leong, and T. Itoh, "Dual-Mode Compact Microstrip Antenna Based on Fundamental Backward Wave," *2005 Asia-Pacific Microwave Conference Proceedings,* Suzhou, China, 2005.

[7] Li, Y., and K. -M. Luk, "A Linearly Polarized Magneto-Electric Dipole with Wide H-Plane Beamwidth," *IEEE Transactions on Antennas and Propagation,* Vol. 62, No. 4, April 2014, pp. 1830–1836.

[8] Ta, S. -X., H. Choo, and I. Park, "A Dual-Band Wide-Beamwidth WLAN Access Point Antenna," *2013 International Workshop on Antenna Technology (iWAT),* Karlsruhe, Germany, 2013, pp. 71–74.

[9] Zhao, L., Y. Li, and J. Wang, "A Wideband Magneto-Electric Dipole Phased Array for 5G Millimeter-Wave Applications," *2020 IEEE MTT-S International Wireless Symposium (IWS),* Shanghai, China, 2020, pp. 1–3.

[10] Liu, N. -W., et al., "A Low Profile Shorted-Patch Antenna with Enhanced Bandwidth and Reduced H-Plane Cross-Polarization," *IEEE Transactions on Antennas and Propagation,* Vol. 66, No. 10, October 2018, pp. 5602–5607.

[11] Liu, N. -W., et al., "Cross-Polarization Reduction of a Shorted Patch Antenna with Broadside Radiation Using a Pair of Open-Ended Stubs," *IEEE Transactions on Antennas and Propagation,* Vol. 68, No. 1, January 2020, pp. 13–20.

3

Multimode Resonance-Inspired Planar Antennas with Impedance Bandwidth Improvement

3.1 Introduction

The low-profile characteristics of microstrip patch antennas lead to several disadvantages, such as high-quality factors, narrow bandwidth, and low efficiency, which restrict their application and development in modern communication systems. Therefore, one important research question is how to control the impedance characteristics of patch antennas or planar antennas to realize a desired wideband or multiband performance. Several methods can be used to achieve a wideband or multiband property for planar antennas: (1) a parasitic element stacked on a principal patch radiator to excite multimode resonance [1]; (2) a feeding network added to realize wide impedance matching [2]; (3) a metamaterial radiation element composed of multiple separated units loaded to achieve the desired broadband bandwidth [3]; (4) modified feeding probe (such as L-shaped, M-shaped, H-shaped, etc.), which leads to the LC (where L represents the inductance and C represents the capacitance) resonance around the dominant mode, expanding the working bandwidth of the antenna [4]; and (5) using a variety of slot-feeding structures as an alternative method for bandwidth enhancement (H-type, I-type, L-type, etc.) [5]. Although these methods can effectively control the impedance characteristics of planar antennas, achieving the working bandwidth of more than 10%, the higher order modes of these antennas are rarely used, leading to high profile, enlarged size, or complex structure.

To address this issue, a multimode resonance idea in a single cavity is seen as a powerful method. In 2005, the ultrawideband filter design theory was proposed based on its fundamental mode and high-order modes [6]. In 2005, the TM_{10} and TM_{01} modes of the microstrip patch antenna were used to achieve low profile, wideband, and broadside radiation patterns [7]. In 2013, the TM_{01} and TM_{02} modes of the circular patch antenna were combined with each other to realize a low profile, wideband, and monopole-like radiation beam [8]. In 2015 and 2017, broadband planar slot antennas based on the first- and third-order resonant modes were proposed, designed, and discussed [9–11]. Since 2015, we have carried out a series of studies on a multimode resonance-inspired method for controlling input impedance, mainly composed of the wide impedance bandwidth techniques in Section 3.2, multi-impedance bandwidth techniques in Section 3.3, and high-order mode-suppression bandwidth techniques in Section 3.4.

3.2 Wide Impedance Bandwidth Techniques

In this section, we discuss the wide impedance bandwidth technique for the multimode resonance-inspired planar antenna, which contains wide impedance bandwidth from the stepped impedance resonator (SIR) technique in Section 3.2.1, shorting pins technique in Section 3.2.2, shorting pins and slots technique in Section 3.2.3, and feeding network technique in Section 3.2.4. The realization process needs to address three critical issues: (1) How to select the effective resonant modes of the planar antennas, (2) how to suppress the undesired modes from the excited ones, and (3) how to tune the frequency ratio between these effective modes.

3.2.1 Stepped Impedance Resonator Technique

Figure 3.1 shows the configuration and fabrication of the dual-mode wideband patch antenna based on SIR in [12], which is mainly composed of the rectangular radiation patch, narrow open stubs, folded ground plane, and differential feed scheme. The rectangular patch with a pair of narrow open stubs at its two radiating edges is combined to form up the SIR. The detail parameters of the antenna are: $L = 45$ mm, $L_1 = 40$ mm, $L_3 = 60.7$ mm, $L_4 = 360$ mm, $W = 150$ mm, $W_1 = 40$ mm, $W_2 = 0.2$ mm, $H = 0.813$ mm, $H_1 = 48$ mm; Y_{stub} is the admittance observed from the radiation patch edge to the open stubs. The dielectric substrate has a relative permittivity of $\varepsilon_r = 3.55$, thickness of $H = 0.813$ mm, and loss tangent of 0.0027.

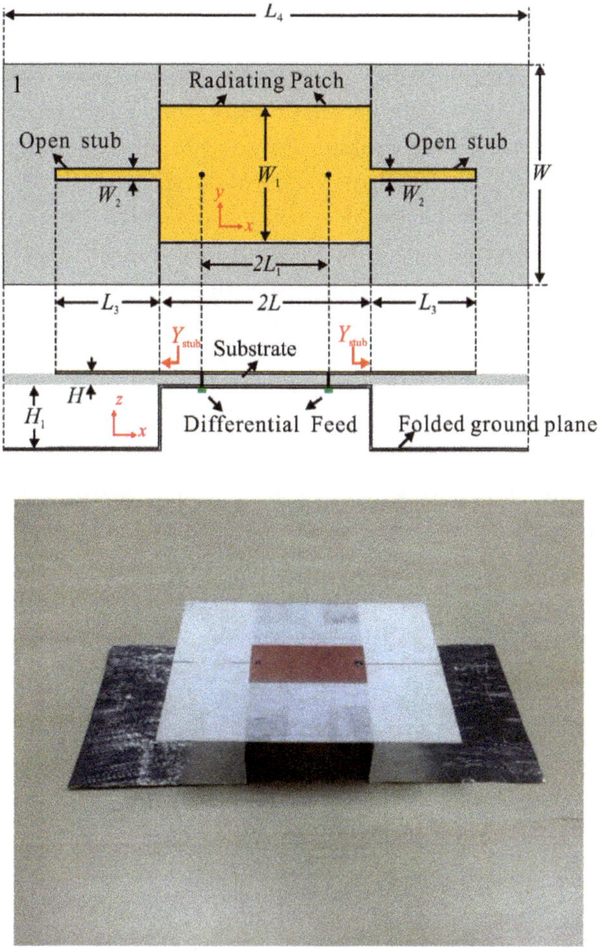

Figure 3.1 Configuration and fabrication of a dual-mode wideband patch antenna based on the SIR technique.

To realize the desired wide bandwidth, we need to address three critical issues. First, the TM_{10} and TM_{30} modes of the patch antenna are selected as the effective modes since they keep the similar broadside radiation patterns. The TM_{01}, TM_{02}, TM_{20}, TM_{12}, TM_{21}, and TM_{22} modes are defined as undesired. Second, the feeding probe is placed in the center of the radiation patch along the y-axis in order to effectively suppress the TM_{01}, TM_{21}, and TM_{22} modes. Third, the differential feeding scheme along the x-axis direction is used to form an equivalent electric wall to suppress the undesired TM_{02} and TM_{20} modes. Next, by reducing the patch width W_1, the resonant frequency of its TM_{12} mode is moved outside to that of TM_{30} mode. Then, by using classical transmission-line models with SIR technology, the resonant frequencies of its TM_{10}

and TM_{30} modes are theoretically studied, and the dual mode could be moved in proximity to each other for impedance bandwidth enhancement.

To investigate the working principle about how to move TM_{10} and TM_{30} modes in proximity to each other, the folded ground structure in Figure 3.1 is replaced by the traditional ground plane to simplify the analysis process, as shown in Figure 3.2. The detail parameters of the antenna are: $L = 45$ mm, $L_1 = 8$ mm, $L_2 = 37$ mm, $L_3 = 48$ mm, $W_1 = 40$ mm, $W_2 = 0.2$ mm, $H = 0.813$ mm.

Meanwhile, in Figure 3.2, the corresponding equivalent transmission-line model is also plotted based on classical transmission-line theory. Here, Y_1 and Y_2 represent the characteristic admittance of the rectangular radiation patch and open stubs, respectively. Their phase constant is about β and β_1; $Y_s = G_s + jB_s$ is the self-admittance on the patch edges; $Y_m = G_m + jB_m$ is the mutual-admittance between these two radiating edges; L_{probe} is the equivalent inductance derived from the feeding probe; $V+$, V_- is the positive and negative voltages on the differential feeding port; and V_1, V_2 is the voltage between the edges of the radiation patch and the ground plane.

Considering the symmetrical differential feeding scheme in Figure 3.2, which keeps equal amplitude and a 180°-phase difference between two ports, the symmetrical plane of the patch center could be regarded as a virtual electric wall. The circuit model in Figure 3.2 can be simplified to Figure 3.3; while Y_{inL} and Y_{inR} are the input admittance from the feeding port as viewed to the left and right sides, respectively.

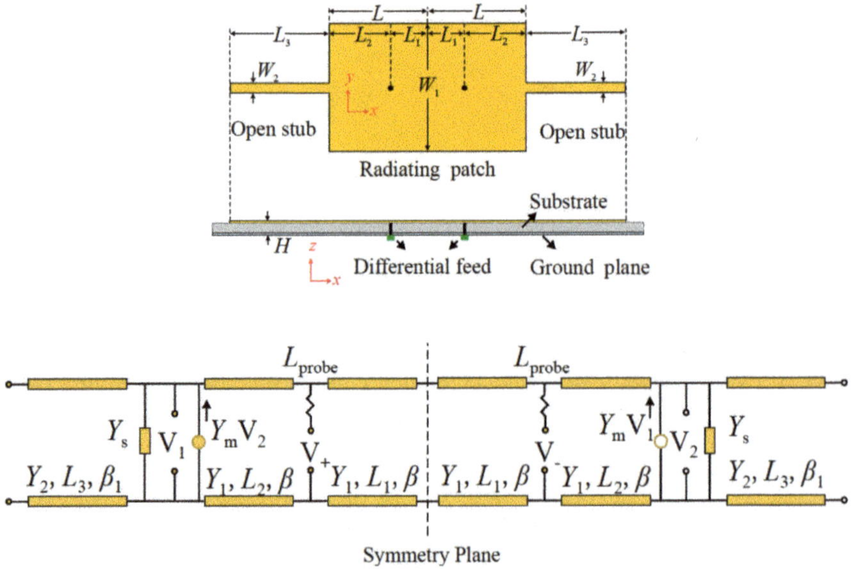

Figure 3.2 Configuration and equivalent transmission-line model of a traditional patch antenna based on the SIR technique.

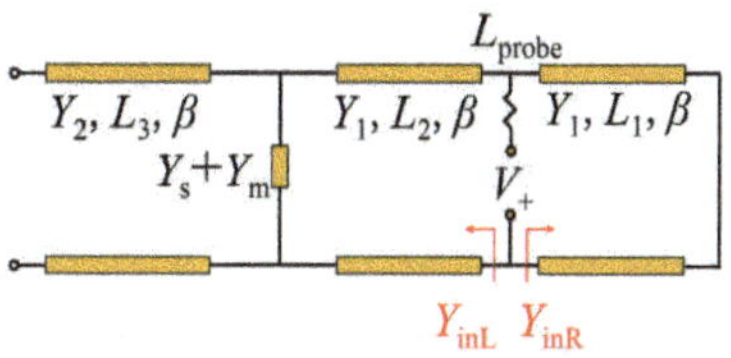

Figure 3.3 Simplified transmission-line model of the antenna from Figure 3.2.

Based on the equivalent model in Figure 3.3, the input admittance Y_{in} can be derived

$$\begin{aligned} Y_{in} &= Y_{inR} + Y_{inL} \\ &= -j*Y_I\cot(\beta L_I) + \frac{Y_s+Y_m+j*Y_2\tan(\beta_I L_3)+j*Y_I\tan(\beta L_2)}{1-\frac{Y_2}{Y_I}\tan(\beta L_2)\tan(\beta_I L_3)+j*\frac{(Y_s+Y_m)}{Y_I}\tan(\beta L_2)} \end{aligned} \quad (3.1)$$

To simplify the calculation process, the L_1, θ, and θ_3 are represented as L, βL, and $\beta_1 L_3$, respectively. Hence, the Y_{in} becomes

$$Y_{in} = G_s+G_m+j*\begin{pmatrix} Y_2\tan(\theta_3)-Y_I\cot(\theta)+Y_I*\tan(\beta*\Delta L) \\ +Y_I*\tan(\beta*\Delta L)*F_b*K_b \end{pmatrix} \quad (3.2)$$

where ΔL is the open-ended extension of the patch, F_b and K_b are the expressions defined in [13].

Considering the resonant condition, the imaginary part of the input admittance Y_{in} should vanish. Hence, the resonant frequencies of the MPA at these odd-order modes could be expressed as

$$Y_2\tan(\theta_3) = Y_I\cot(\theta)\text{-}B_s\text{-}B_m \quad (3.3)$$

From (3.3), we see that the resonant frequencies of TM_{30} and TM_{10} modes are mainly determined by two key parameters: SIR impedance ratio $R_z = Y_1/Y_2$, SIR electrical length ratio $u = \theta_3/\theta$. By increasing the value of R_z, the dual-mode frequency ratio f_3/f_1 could be decreased, and the smallest f_3/f_1 could be achieved by setting u as 1.

Due to fabrication limitations, the maximum impedance ratio R_z only reaches to about 32 in a single-layer implementation in Figure 3.2, which still keeps a large frequency ratio between these two modes. It is important to fold the ground plane as the stepped scheme to further increase the value of R_z, as observed in Figure 3.1, and the corresponding fabricated antenna with simulated and measured results are plotted in Figure 3.4. Results show that the

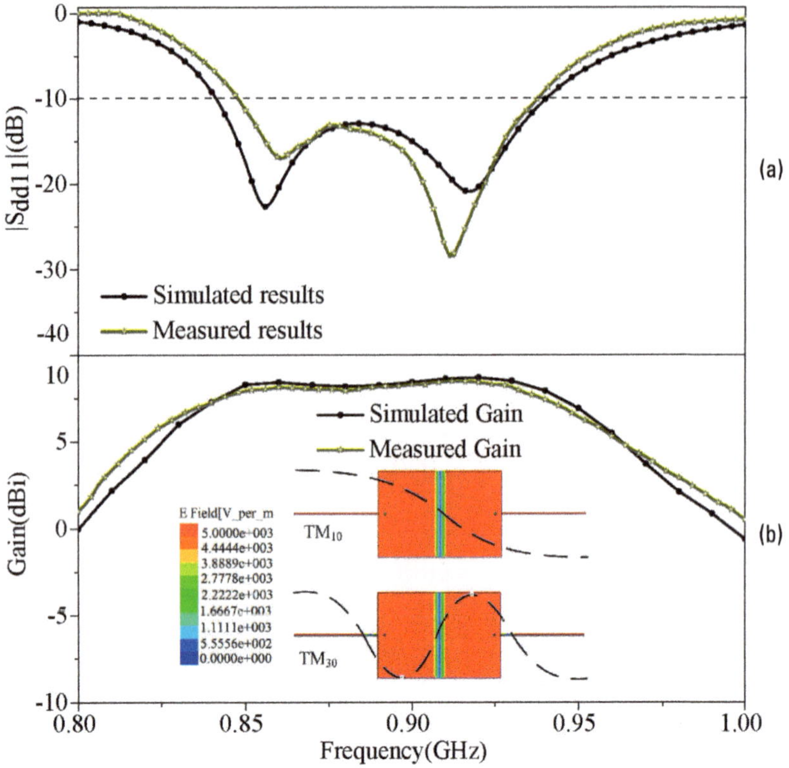

Figure 3.4 Simulated and measured results of a dual-mode wideband patch antenna based on the SIR: (a) S-parameter, and (b) gain and electric-field distribution.

measured impedance bandwidth reaches about 10% ranging from 0.85 to 0.94 GHz. Meanwhile, a stable gain between 7.4 and 8.5 dB is achieved during the wide working band. The electric-field distribution also verifies that the TM_{10} and TM_{30} modes are successfully excited for the designed antenna at lower and upper frequency, respectively.

The differential feeding scheme is used to suppress undesired even-order modes of the antenna. Meanwhile, by decreasing the width of the radiator, the resonant frequency of its TM_{12} mode could be tuned outside to the resonant frequencies between its TM_{10} and TM_{30} modes. In addition, the SIR technology could be used to reallocate resonant frequencies of its TM_{10} mode and TM_{30} mode in proximity to each other. Furthermore, the folded ground structure brings the TM_{10} and TM_{30} modes closer. Finally, the designed multimode patch antenna has the characteristics of wide bandwidth, stable radiation gain, high efficiency, and low cross polarization.

3.2.2 Shorting Pins Technique

The work in Section 3.2.1 realizes wide bandwidth performance at dual modes at the expense of antenna profile. In this part, we will discuss a low-profile wideband microstrip patch antenna based on the loaded shorting pins [14], which still combines TM_{10} and TM_{30} modes of the patch antenna.

Figure 3.5 shows the configuration and fabrication of a low-profile dual-mode wideband MPA based on the loaded six shorting pins, composed of the radiation patch, shorting pins, microstrip feeding, ground plane, slot, dielectric substrate 1, dielectric substrate 2, and so on. The specific parameters of the antenna are: L = 140 mm, L_1 = 200 mm, L_2 = 100 mm, L_{slot} = 31 mm, L_s = 8.15 mm, L_{feed} = 9.85 mm, W = 50 mm, W_{feed} = 1.82 mm, W_{slot} = 0.3 mm, R = 2 mm, S_1 = 98 mm, S_2 = 32.5 mm, H = 1.8 mm. The relative permittivity and thickness of the dielectric substrate 1 are 1.96 and 1.27 mm. Meanwhile, the relative permittivity and thickness of the dielectric substrate 2 are 3.38 and 0.813 mm.

First, we still select TM_{10} and TM_{30} modes as effective modes to realize the broadside radiation pattern in the wideband, such that the undesired modes should be TM_{01}, TM_{02}, TM_{20}, TM_{12}, and TM_{21} modes. Second, the virtual

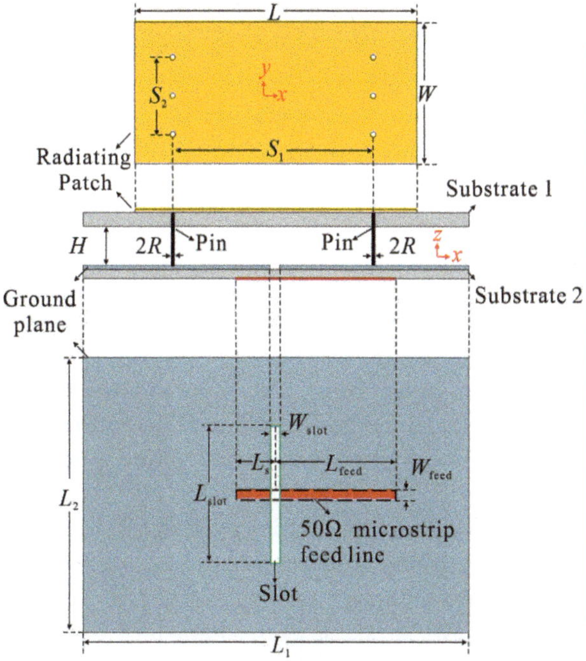

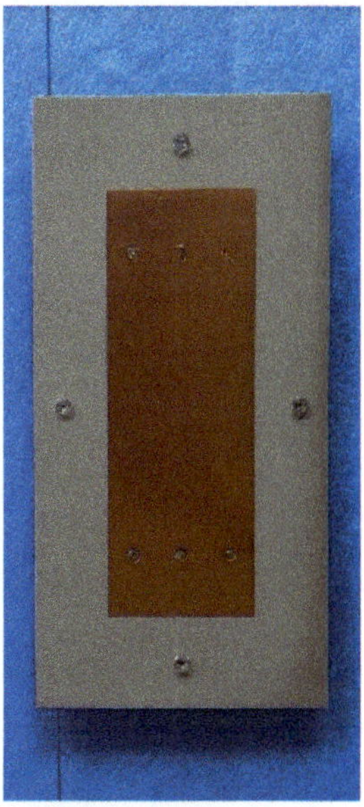

Figure 3.5 Configuration and fabrication of a low-profile dual-mode wideband microstrip patch antenna based on the loaded shorting pin technique.

electric wall is formed by the aperture-coupled microstrip feeding, suppressing these even-order modes of the antenna, such as TM_{01}, TM_{02}, TM_{20}, and TM_{21} modes. Third, the width W of the radiation patch is reduced to less than 0.6 times the length L, so that the resonant frequency of its TM_{12} mode is moved outside to that of TM_{10} mode and TM_{30} mode. Fourth, the shorting pins are used to disturb the electric-field distribution characteristics of the dual mode for frequency reallocation.

On one hand, the shorting pins should be loaded around the electric-field antinode of its TM_{10} mode, thereby changing the inherent electric-field amplitude and increasing its resonant frequency. On the other hand, the shorting pins should be loaded around the electric-field node of its TM_{30} mode, which slightly perturbs the original electric field and keeps its stable resonant frequency.

Based on the above principle, the impedance characteristics of microstrip patch antennas with different numbers of shorting pins at TM_{10} and TM_{30} modes are discussed in Figure 3.6. When the shorting pins are unloaded, the resonant frequency ratio between TM_{30} mode and TM_{10} mode is maintained around 3, as shown in Figure 3.6(a). With an increasing number of shorting pins, the resonant frequency of TM_{10} mode is significantly tuned up, while the resonant frequency of TM_{30} mode keeps almost constant in Figure 3.6(b–d). When the pins' number is increased to six, the resonant frequencies of TM_{30} and TM_{10} modes are moved in proximity to each other, forming a low-profile wideband performance as shown in Figure 3.6(d).

The antenna configuration in Figure 3.5 shows the proposed antenna, simulated and measured. The simulation and test S-parameter and gain results of this antenna are plotted in Figure 3.7. The antenna maintains wideband characteristics in the operating frequency band range from 2.32 to 2.70 GHz (15.1%), and the profile of the antenna is as small as about 0.032 λ_0 (λ_0 is the free-space wavelength at the center frequency). The gain of the proposed antenna in wideband is maintained between 3 and 6.8 dBi. Although both modes have been perturbed, they demonstrate the basic property of TM_{10} and TM_{30} modes.

The aperture-coupled feeding scheme is used to suppress resonant frequencies of the antenna under even-order modes. The width of the core radiator is reduced to tune the resonant frequency of TM_{12} mode out of the bandwidth composed by TM_{10} and TM_{30} modes. Additionally, shorting pins are loaded around the electric-field node of TM_{30} mode, increasing the resonant frequency of TM_{10} mode while keeping the relatively stable resonant frequency of TM_{30} mode. Finally, the designed dual-mode resonant antenna has the characteristics of low profile, wide impedance bandwidth, stable gain, and high efficiency.

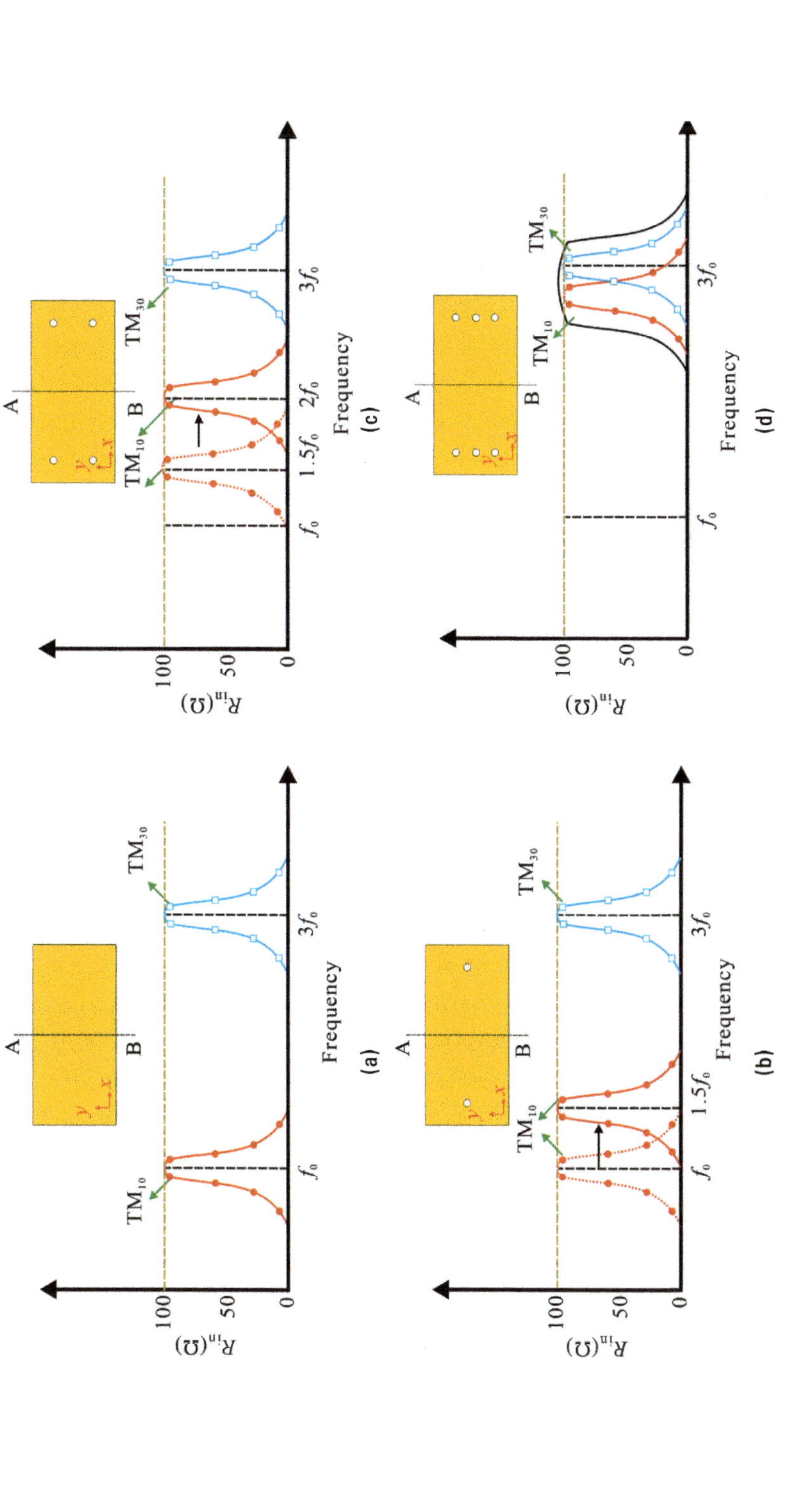

Figure 3.6 Input resistance variation of the antenna at TM_{10} and TM_{30} modes under different number of shorting pins: (a) no pins, (b) two pins, (c) four pins, and (d) six pins.

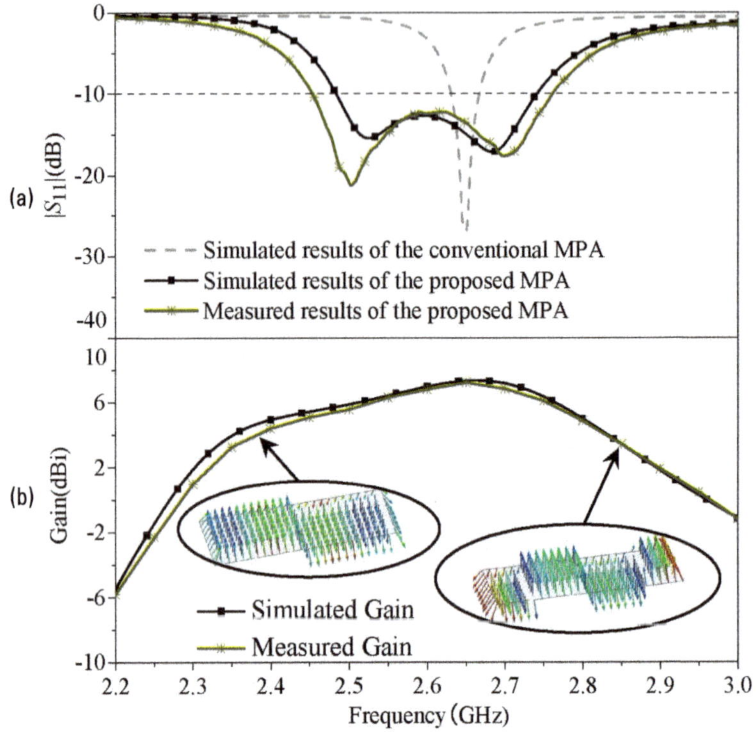

Figure 3.7 The simulated and measured results of the proposed dual-mode wideband microstrip patch antenna based on six shorting pins: (a) S-parameter, and (b) gain and internal electric-field distribution.

3.2.3 Shorting Pins and Slots Techniques

In Section 3.2.2, although the dual-mode antenna achieves wide bandwidth and low profile simultaneously by loading the shorting pins, the dimension of the antenna is significantly enlarged due to the increased resonant frequency. This part will discuss a miniaturized method for the multimode wideband microstrip patch antenna based on the shorting pins and slot [15], which mainly incorporates the $TM_{0,1/2}$ and $TM_{0,3/2}$ modes of the shorted patch antenna (SPA).

The configuration and fabrication of the antenna is depicted in Figure 3.8; it is composed of the radiation patch, shorting wall, V-shaped gap, shorting pins, dielectric substrate, and ground plane. The detail parameters of this designed antenna are: $L = 25$ mm, $L_1 = 27$ mm, $L_g = 60$ mm, $W = 30$ mm, $W_1 = 0.8$ mm, $W_g = 35$ mm, $D = 5.8$ mm, $D_1 = 15$ mm, $D_2 = 18.75$ mm, $S_1 = 19.5$ mm, $R = 1.2$ mm, $\alpha = 160$. The relative permittivity of the dielectric substrate is 2.2 and its thickness is 3.18 mm.

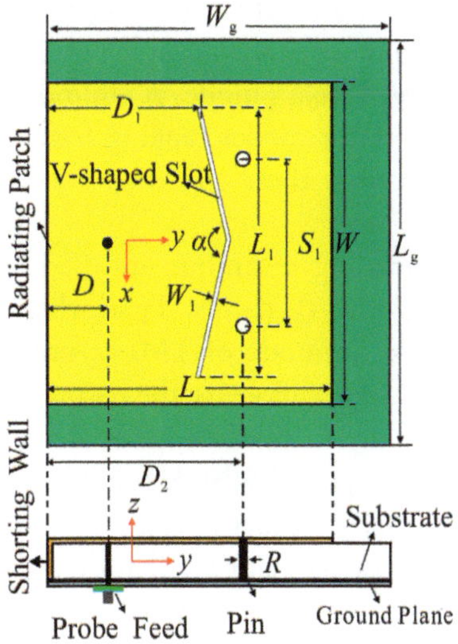

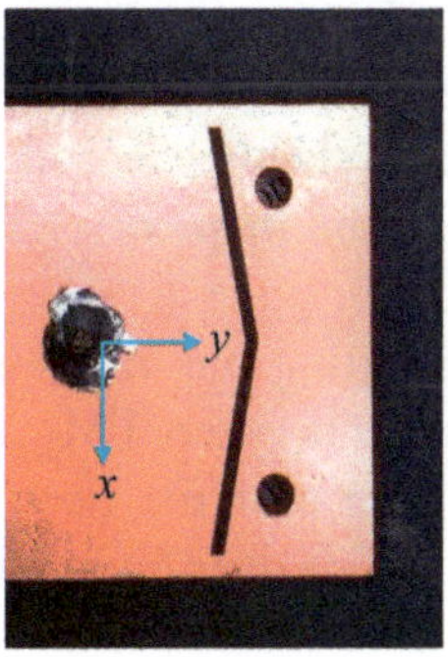

Figure 3.8 Schematic diagram and fabrication of a miniaturized dual-mode wideband microstrip patch antenna based on the loaded shorting pins and slot techniques.

Based on the cavity-mode theory, the resonant frequency f_{mn} and wavenumber k_{mn} of the traditional patch antenna under TM_{mn} mode could be expressed as

$$f_{mn} = c\sqrt{(m\pi / W_e)^2 + (n\pi / L_e)^2} / (2\pi\sqrt{\varepsilon_r}) \tag{3.4}$$

where W_e and L_e are the effective width and length of the rectangular radiation patch by considering its edge influence, respectively, c is the light speed in vacuum, m and n are equal to 0, 1, 2.

With these odd-order modes, the symmetrical plane of the rectangular patch is a perfect electric wall. The shorting wall is introduced in this symmetrical plane to form up the conventional SPA in Figure 3.8, and only the half-odd-order modes of the antenna could be excited and operated for radiation. Therefore, all even-order modes of a traditional MPA are successfully suppressed. Most important, antenna size is also significantly reduced. Since the resonant frequency of the SPA at $TM_{2,1/2}$ mode is between its $TM_{0,1/2}$ and $TM_{0,3/2}$ modes, it is important to adjust the width and length of the radiating patch. Accordingly, W_e and L_e should satisfy the following requirement to move the resonant frequency of its $TM_{2,1/2}$ mode outside to that of $TM_{0,1/2}$ and $TM_{0,3/2}$ modes

$$(2\pi / W_e)^2 + (\pi / 2L_e)^2 > (3\pi / 2L_e)^2 \tag{3.5}$$

By properly selecting the width and length of the radiating patch based on (3.5), the input resistance of the antenna is shown in Figure 3.9(a). The traditional antenna resonates $TM_{0,1/2}$ and $TM_{0,3/2}$ modes at around 1.86 and 5.62 GHz, respectively, with no other resonant modes ranging from 1 to 6 GHz. Nevertheless, the frequency ratio between these two modes is kept relatively large at about 3.

The shorting pin and V-slot loading technique are used to reduce the frequency ratio between $TM_{0,1/2}$ and $TM_{0,3/2}$ modes. First, when the shorting pins are loaded near the electric-field antinode of $TM_{0,1/2}$ mode, where the electric-field node distribution of its $TM_{0,3/2}$ mode is, the resonant frequency of $TM_{0,1/2}$ mode could be effectively increased up, while the resonant frequency of $TM_{0,3/2}$ mode is slightly influenced in Figure 3.9(b). Second, by cutting the V-shaped slot around this place, the resonant frequency of $TM_{0,3/2}$ mode could be effectively reduced while keeping almost constant for that of $TM_{0,1/2}$ mode in Figure 3.9(c). The V-shaped slot instead of the linear slot is adopted to improve its impedance matching over the wide operation band.

The proposed antenna is designed, fabricated, and measured in Figure 3.8. The simulation and measured S-parameter results of this antenna are shown in Figure 3.10. The simulated and measured results match well,

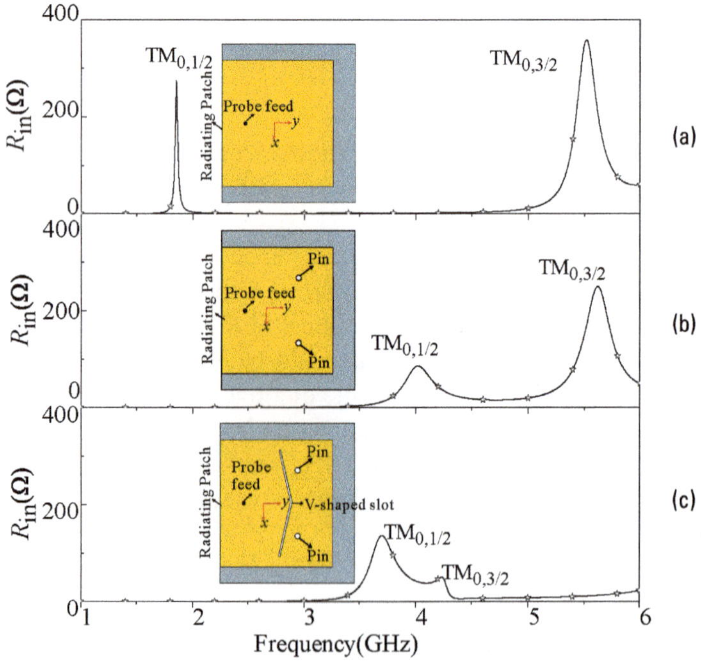

Figure 3.9 Input resistance curves of the SPA under different pins and slots: (a) traditional SPA, (b) traditional SPA with loaded pins, and (c) traditional SPA with loaded pins and slot.

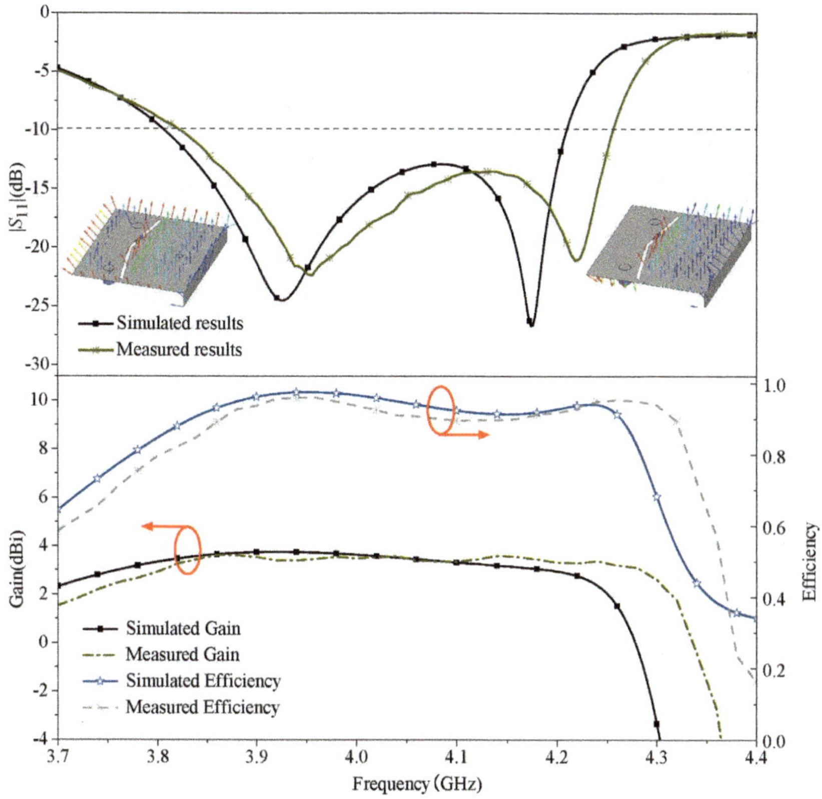

Figure 3.10 Simulated and measured results of the fabricated compact dual-mode wideband SPA based on shorting pins and V-shaped slots: (a) S-parameter, and (b) gain and efficiency.

evidently confirming the predicted performances of these two antennas. The measured impedance bandwidth for $|S_{11}|$<-10 dB is kept at about 11.8% within the frequency range of 3.85 to 4.33 GHz, while that of the traditional SPA only gains the impedance bandwidth of about 5.8% within the range of 4.03 to 4.27 GHz. The antenna obtains a reasonably stable gain of around 3.0 dBi at the broadside direction in the entire operating band, while keeping a high efficiency of around 85%. Moreover, the electric-field distribution proves that the electric field has no transverse phase reversal at 3.95 GHz. The $TM_{0,1/2}$ mode is excited in the lower band, but the transverse electric field varies once at 4.25 GHz, thus representing the operation of the $TM_{0,3/2}$ mode in the upper band.

The shorting wall is used to reduce the electrical size of the antenna. The shorting pin is loaded to improve the resonant frequency of $TM_{0,1/2}$ mode. The V-shaped slot is used to reduce the resonant frequency of $TM_{0,3/2}$ mode and electrical size of the antenna. Finally, the designed dual-mode resonant SPA has the attractive features of low profile, miniaturization, wide band, stable gain, and high efficiency.

3.2.4 Feeding Network Technique

Adding a feeding network is an alternate method for impedance bandwidth enhancement. In this part, we will design a low-profile wideband patch antenna under quintuple-mode resonance by introducing a feeding network. Here, the TM_{10} mode, TM_{12} mode, TM_{30} mode, feeding network SRR1, and feeding network SRR2 are combined for impedance-bandwidth enhancement.

Figure 3.11 shows the antenna configuration detail, composed of the radiating patch, shorting pins, feeding network, aperture-coupled feeding, dielectric substrate 1, dielectric substrate 2, and so on. The feeding network consists of two split-ring resonators (SRR) and a parallel coupled line. The detail parameters of this designed antenna are: $L_1 = 100$ mm, $L_2 = 98$ mm, $L_3 = 6.5$ mm, $L_4 = 12$ mm, $L_5 = 42.5$ mm, $L_6 = 11.75$ mm, $L_7 = 10.7$ mm, $S_1 = 44.5$ mm, $S_2 = 42.4$ mm, $S_3 = 0.2$ mm, $S_4 = 0.25$ mm, $g_1 = 2.5$ mm, $g_2 = 0.3$ mm, $H = 4$ mm, $H_1 = 1$ mm, $H_2 = 0.8$ mm, $R_1 = 0.5$ mm, $d_1 = 2.2$ mm, $W_1 = 52$ mm, $W_2 = 5.5$ mm, $W_3 = 3.12$ mm, $W_4 = 0.5$ mm, $G_1 = 170$ mm, $G_2 = 70$ mm, $l = 11$ mm, $l\_s = 34$ mm, $w\_s = 0.5$ mm. The relative permittivity of the dielectric substrate 1 is 2.2 and its thickness is 0.8 mm. The relative permittivity of the dielectric substrate 2 is 2.2 and its thickness is 1 mm [16].

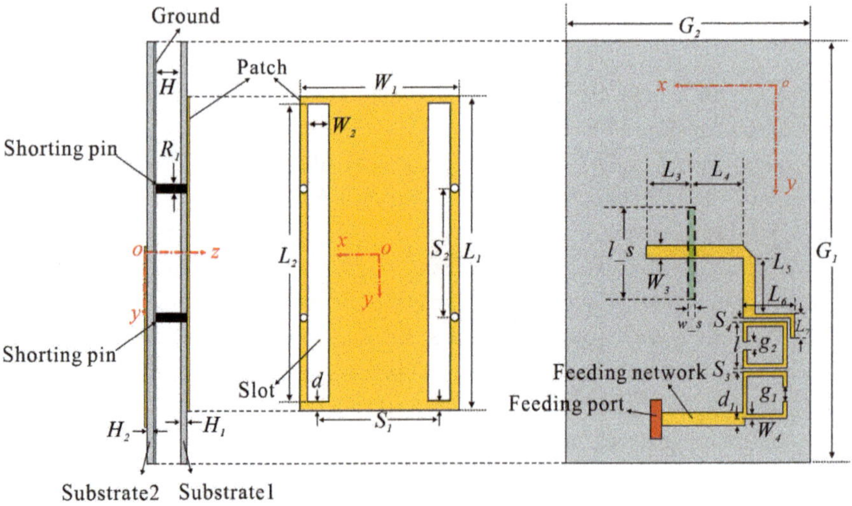

Figure 3.11 The configuration of a wideband patch antenna under quintuple-mode resonances.

Figure 3.12 plots the input resistance variation of the proposed antenna with a feeding network. The antenna successfully resonates the desired TM_{10}, TM_{12}, and TM_{30} modes at around 2.85, 2.65, and 2.33 GHz, respectively. Meanwhile, the SRR is also introduced to excite the SRR1, SRR2 modes from the feeding network, which is kept in proximity to that of TM_{10}, TM_{12}, and TM_{30} modes. Under this condition, the impedance bandwidth of the patch antenna is further extended under quintuple-mode resonance.

Figure 3.13 shows the detail configuration and equivalent circuit of the SRR, which is composed of the SRR1, SRR2, and the parallel-coupled microstrip feeding line. Here the SRR1 from the feeding network is represented by using L_1 and C_1; the SRR2 from the feeding network is represented by using L_2 and C_2.

The proposed antenna is designed and simulated in Figure 3.11, and the simulated S-parameters, gain, and efficiency of this antenna are shown in Figure 3.14. The impedance bandwidth for $|S_{11}| < -10$ dB is equal to about 26% with a range of 2.29 to 2.96 GHz for the proposed antenna, while keeping the low-profile performance of 0.042 λ_0. Meanwhile, the simulated gain and efficiency

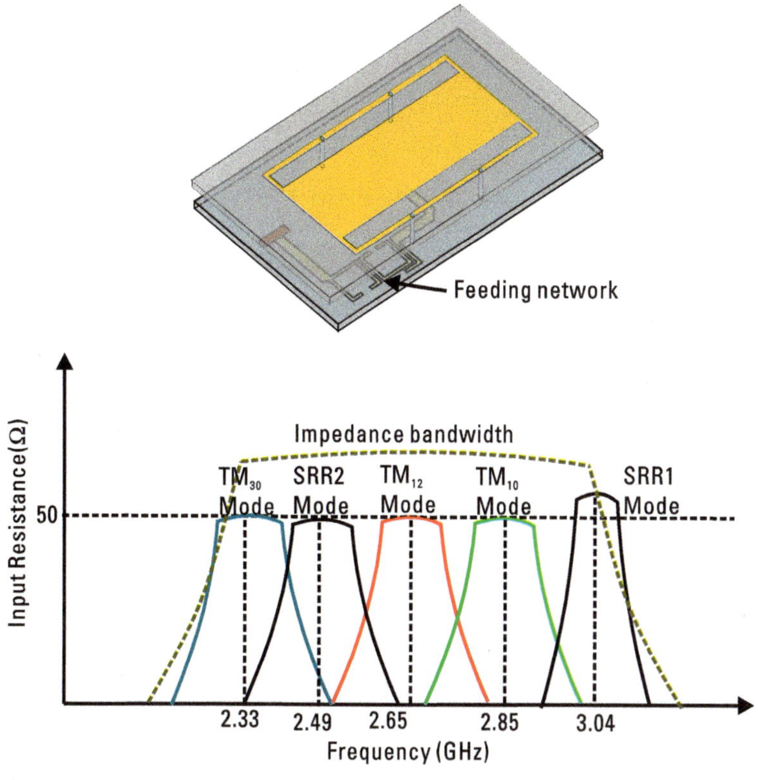

Figure 3.12 Resistance characteristics variation of the antenna with the feeding network.

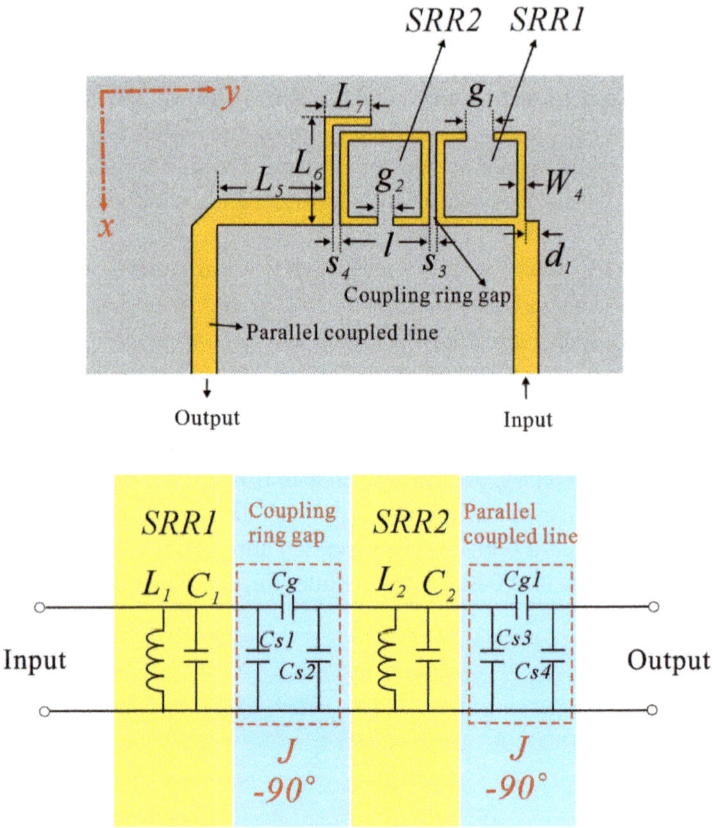

Figure 3.13 Detail dimensions and equivalent circuit of the feeding network.

of the antenna are maintained at around 7.0 dBi and 90%, keeping the stable radiation pattern and high efficiency.

The aperture-coupled feeding scheme is used to suppress even-order modes. The shorting pins and slot are used to reallocate TM_{10}, TM_{12}, and TM_{30} modes in proximity to each other. In addition, the SRR1 and SRR2 modes are excited by using the feeding network. Finally, the low profile, wide bandwidth, stable gain, and low cross polarization are simultaneously achieved for the patch antenna. Previous works have been proposed and developed using the multimode concept shown in Figure 3.15. In [9], the first- and second-odd-order modes could be shifted and combined by using the stub-loaded slot. In [18, 20] the pin-loaded element is used to reallocate its fundamental mode for dual-band generation. As reported in [21], the U-slot and shorting pins are introduced to excite nonresonant mode for bandwidth enhancement.

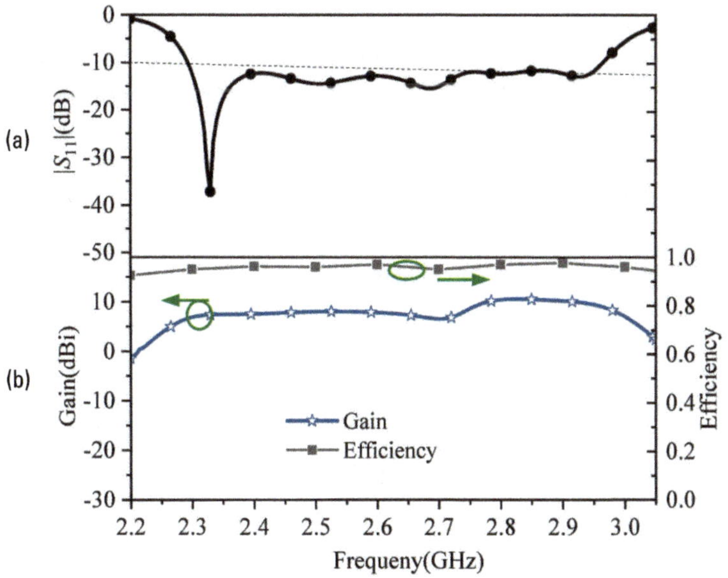

Figure 3.14 Simulated results of the designed MPA with the feeding network in Figure 3.11: (a) $|S_{11}|$, and (b) gain and efficiency.

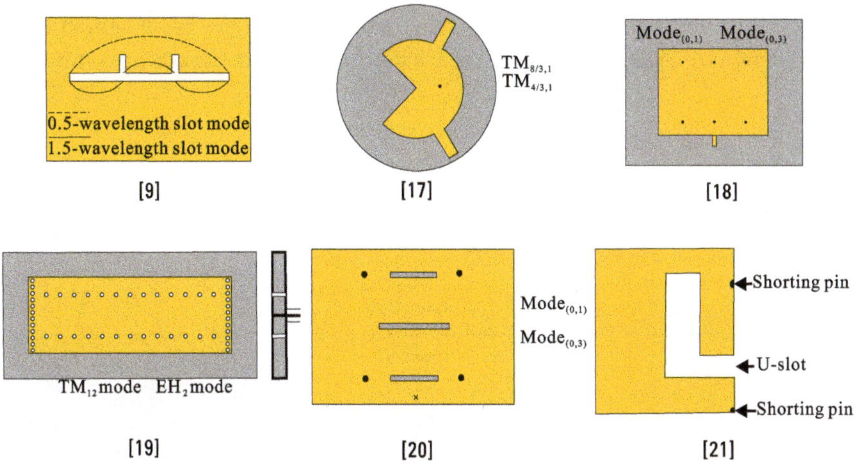

Figure 3.15 The previous antennas with wideband or multiband based on the multimode inspired concept.

3.3 Multi-Impedance Bandwidth Techniques

Multimodes could be combined for wide bandwidth and could be independently controlled for multiband resonance. In this part, we discuss the multi-impedance bandwidth technique for the multimode resonance-inspired planar antenna. It contains the multiband LP antenna in Section 3.3.1 and the multiband circularly polarized (CP) antenna in Section 3.3.2. Here, the realization process needs to address three critical issues: (1) how to select these effective modes of the patch antenna, (2) how to remove these undesired modes between the effective modes, and (3) how to independently control the resonant frequency of these desired modes.

3.3.1 Multiband LP Antenna Technique

For an LP antenna, it is easy to excite the fundamental and higher order modes for the generation of multiband performance. As reported in [22], by employing TM_{10}, TM_{12}, and TM_{30} modes simultaneously, the MPA operating in triple WLAN bands (2.4/5.2/5.8 GHz) is realized, while keeping the low profile, stable broadside radiation, and stable gain as desired. The working principle about how to realize this performance will not be discussed in this part.

3.3.2 Multiband CP Antenna Technique

Multiband CP performance could be also realized by combining several radiative modes of a single radiator. In this part, the realization method of the multifrequency CP antenna based on the multimode theory is extensively studied [23], noticing that the TM_{10}, TM_{01}, and TM_{20} modes of the antenna are simultaneously excited and combined. Figure 3.16 gives the configuration and fabrication of the proposed antenna, which is mainly composed of the radiation patch, shorting pins, open-ended stub, dielectric substrate, feeding probe, and ground plane. The detail parameters of the antenna are: $L_g = 110$ mm, $L_1 = 50$ mm, $L_2 = 8.1$ mm, $L_3 = 21.7$ mm, $D_1 = 16.25$ mm, $D_2 = 13$ mm, $W_1 = 43.1$ mm, $W_2 = 1.1$ mm, $S_1 = 2.5$ mm, $S_2 = 1.5$ mm, $R = 0.9$ mm, $H = 3$ mm. The

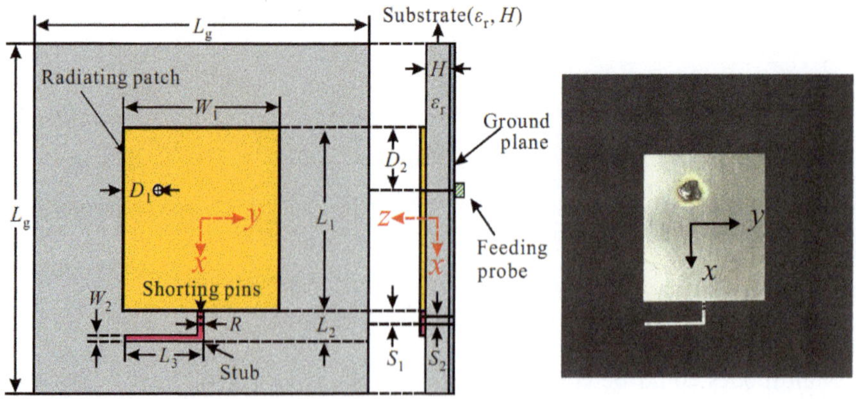

Figure 3.16 Configuration and fabrication of a dual-band CP microstrip patch antenna with extremely small frequency ratio based on triple-mode resonance.

relative permittivity of the dielectric substrate is 2.1 with a thickness of 3 mm. The designed antenna has several attractive characteristics: low profile, small size, left-handed circular polarization (LHCP) pattern, right-handed circular polarization (RHCP) pattern, extremely small frequency ratio, and high efficiency. The detailed design process is discussed in the following.

The first step is to analyze the equivalent magnetic current distribution of the antenna under TM_{10} mode, TM_{01} mode, and TM_{20} mode. Figure 3.17 gives the electric-field and equivalent magnetic-current (EMCs) distribution

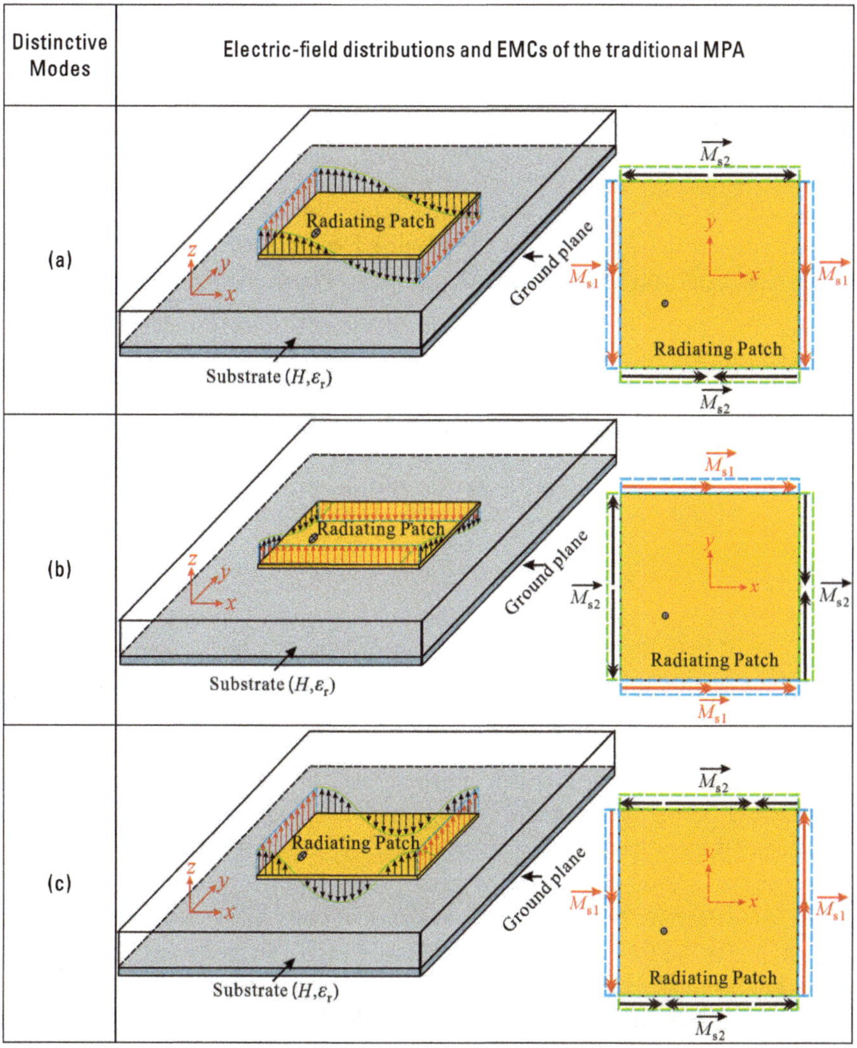

Figure 3.17 Electric-field distributions and EMCs of a traditional MPA under three distinctive modes: (a) TM_{10} mode, (b) TM_{01} mode, and (c) TM_{20} mode.

diagram of the traditional patch antenna under different modes. When the antenna is resonated at TM_{10} mode, the red EMC $\overrightarrow{M_{s1}}$ along the y-axis direction generates the main radiation fields. When the antenna resonates in the TM_{01} mode, the red EMC $\overrightarrow{M_{s1}}$ along the x-axis direction produces the main radiation fields. When the antenna is operated at TM_{20} mode, the red EMC $\overrightarrow{M_{s1}}$ along the y-axis produces the main radiation fields. Unfortunately, the red EMC $\overrightarrow{M_{s1}}$ of the antenna at TM_{20} mode presents the opposite direction, while the red EMC of the antenna at TM_{10} and TM_{01} modes presents the same direction.

Assuming that the electric fields across these red EMCs are uniformly distributed, the far-zone normalized patterns of TM_{10} mode in Figure 3.17(a) at xoz-plane could be deduced as

$$\begin{cases} E_\theta = \cos(0.5k_0 L_e \sin\theta) \\ E_\varphi = 0 \end{cases} \tag{3.6}$$

where L_e is the electrical length by considering the patch edge effect.

By using the red EMC, the normalized xoz-plane radiation fields of the microstrip patch antenna under TM_{01} mode in Figure 3.17(b) can be expressed as

$$\begin{cases} E_\theta = 0 \\ E_\phi = \cos(0.5k_0 L_e \sin\theta) \end{cases} \tag{3.7}$$

In addition, based on the red EMC, the normalized xoz-plane radiation fields of the microstrip patch antenna under TM_{20} mode in Figure 3.17(c) can be expressed as

$$\begin{cases} E_\theta = \sin(0.5k_0 L_e \sin\theta) \\ E_\varphi = 0 \end{cases} \tag{3.8}$$

From (3.6) to (3.8), it can be seen that the traditional MPA produces θ-polarized components under TM_{10} mode, $|E_\phi|$ components under TM_{01} mode, and ϕ-polarized components under TM_{20} mode, thus proving that these three modes are orthogonal with each other. Unfortunately, the traditional TM_{20} mode cannot be well applied to the dual-band CP antennas, whose pattern needs to be transformed from the dual radiation beam into the normal radiation beam as desired.

In the second step, to reshape the radiation pattern of TM_{20} mode, a purple EMC $\overrightarrow{M_{s3}}$ is introduced on the right edge of the radiation patch in Figure 3.18, and its direction is opposite to that of the red EMC $\overrightarrow{M_{s1}}$ on the same

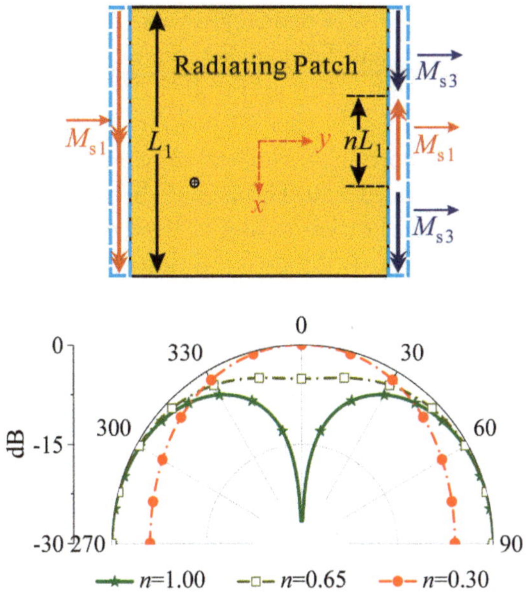

Figure 3.18 Calculated *yz*-plane pattern of a traditional microstrip patch antenna under magnitude ratio *n*.

side. At the same time, we set the magnitude ratio between the right EMC $\overrightarrow{M_{s1}}$ and the left EMC $\overrightarrow{M_{s1}}$ as *n*. Under this condition, the modified radiation field of the MPA under TM_{20} mode resonance is expressed as

$$\begin{cases} E_\theta = \cos(\dfrac{k_0 L_e \sin\theta}{2}) - ne^{j\frac{k_0 L_e \sin\theta}{2}} \\ E_\varphi = 0 \end{cases} \quad (3.9)$$

Based on (3.9), Figure 3.18 also plots the *yz* plane radiation pattern of the antenna under the modified TM_{20} mode as a function of *n*. It can be seen from the figure that when $n = 1$, the EMC $\overrightarrow{M_{s1}}$ on the right side and the EMC $\overrightarrow{M_{s1}}$ on the left side keep the same amplitude and opposite direction, generating the null in the broadside direction. When the value of *n* is gradually decreased, the EMC $\overrightarrow{M_{s1}}$ on the right side also gradually decreases, which increases the magnitude of the radiation pattern at TM_{20} mode in the broadside direction. When $n = 0.3$, the radiation pattern of TM_{20} mode is successfully reshaped into a normal radiation pattern.

In the third step, to control the magnitude ratio *n*, the open-ended stub is loaded around the edge of the radiating patch in Figure 3.16. The red EMCs between TM_{01} and TM_{10} modes rotate clockwise. By contrast, the red EMCs between its reshaped TM_{20} and TM_{01} modes rotate counterclockwise. The dual circularly polarized pattern is generated for the MPA in dual bands by sequentially moving them close to each other in resonant frequencies, which could be realized by loading two shorting pins around the connected place of the radiating patch and the open-ended stub. This method brings an extremely small frequency ratio compared to the traditional multilayer or multiradiator methods.

The proposed antenna is designed, fabricated, and measured in Figure 3.16. The simulation and test S-parameter, axial ratio and gain results of this antenna are shown in Figure 3.19. The designed microstrip patch antenna achieves dual-band performances (2.13 to 2.15 GHz, 2.17 to 2.24 GHz), while keeping the low-profile property of 0.029 λ_0. Meanwhile, this designed antenna forms an LHCP pattern near the lower frequency at 2.134 GHz, and a RHCP pattern near the upper frequency at 2.223 GHz, resulting in an extremely small frequency ratio of about 1.04, while traditional dual-band dual-polarized antennas usually use four radiative modes, and the frequency ratio between these dual bands is only reduced to about 1.1. In addition, the gain of the designed antenna in the working frequency band is maintained at around 7.5 dBic.

The nonbroadside radiation pattern of TM_{20} mode is successfully reshaped into broadside radiation by using an open-ended stub. The TM_{10}, TM_{01}, and TM_{20} modes of the antenna are reallocated in proximity to each other by loading the shorting pins. The proposed multimode resonant antenna has several attractive performances, including low profile, dual band, LHCP radiation pattern, RHCP radiation pattern, extremely small frequency ratio, and high efficiency. Previous works have been proposed and developed by using the multimode concept as shown in Figure 3.20. In [24], four T-shaped slits are loaded around the patch edges with a single feed, thus generating dual-band CP performance under operation of TM_{10} and TM_{30} modes. In [25], four V-shaped asymmetrical resonators, which are placed on a circular substrate symmetrically, are used and designed to generate both RHCP and LHCP performance at two working bands, independently.

3.4 High-Order Suppression Bandwidth Techniques

High-order suppression performance in an antenna helps prevent a communication system from emitting or receiving unnecessary electromagnetic waves. In this part, the radiative modes of the antenna at its higher modes will be discussed and suppressed, which contains a LP antenna with high-order suppression in Section 3.4.1, and a CP antenna with high-order suppression in Section 3.4.2.

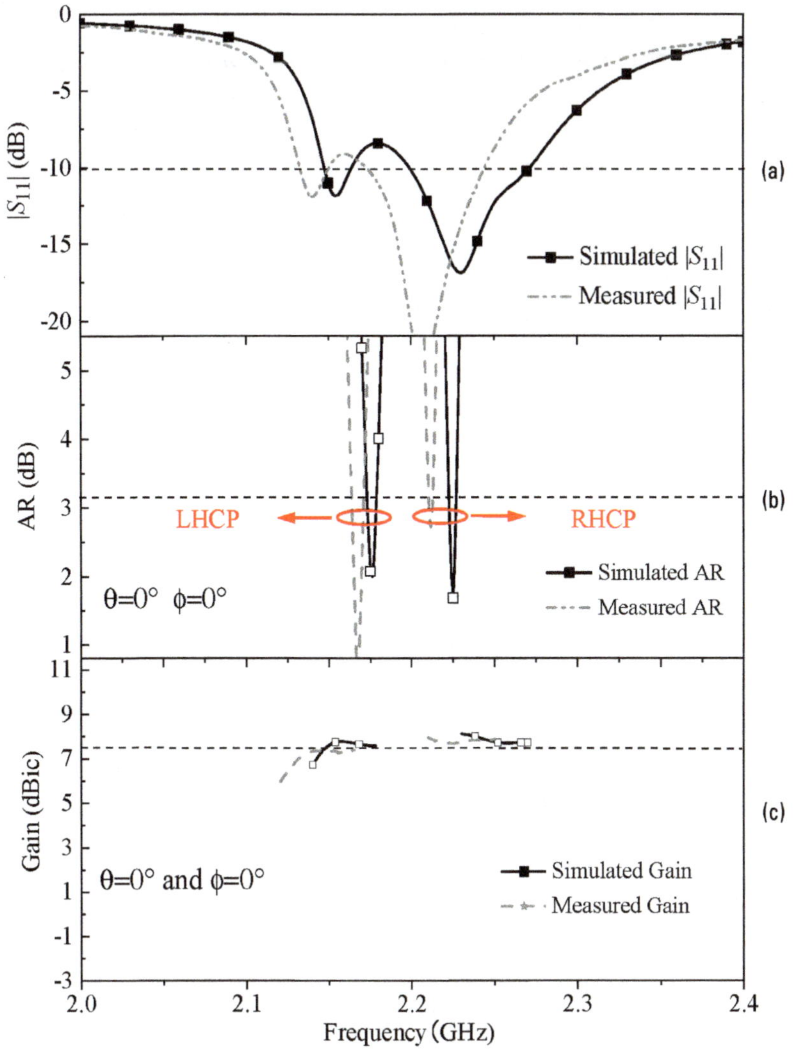

Figure 3.19 Simulated and measured results of the dual-band CP antenna with extremely small frequency ratio based on triple-mode resonance: (a) S-parameter, (b) axial-ratio, and (c) gain.

3.4.1 LP Antenna with High-Order Suppression Technique

This section discusses a design method for wideband harmonic suppression in an LP antenna based on a pair of $\lambda/4$ microstrip lines, which also increase impedance bandwidth under dual resonance. Figure 3.21 is the configuration of the low-profile multimode microstrip patch antenna with high-order suppression [26], which is composed of the radiating patch, T-shaped microstrip line, dielectric substrate, shorting pin, and ground plane. The specific parameters of

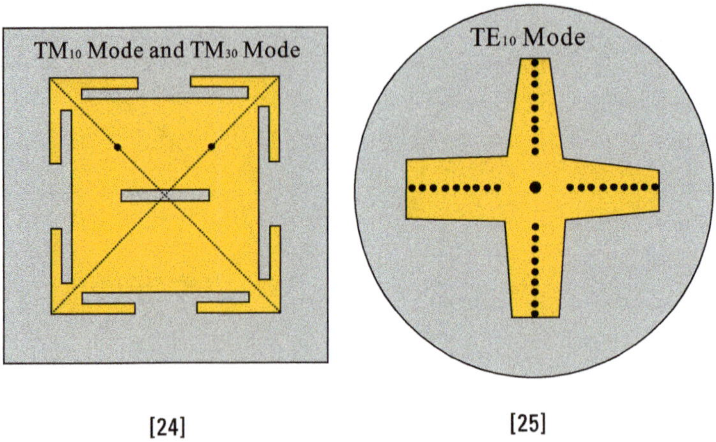

Figure 3.20 CP antennas based on the multimode-inspired concept.

the antenna are: L_p = 18 mm, W_p = 27 mm, L_r = 11.25 mm, W_r = 0.5 mm, d = 1.75 mm, W = 4.5 mm, r = 0.5 mm. The dielectric substrate has a relative permittivity of 2.33 and a thickness of 1.57 mm.

Meanwhile, Figure 3.21 shows the equivalent circuit of a low-profile multimode microstrip patch antenna with high-order suppression, where the radiation patch is equivalent to R_1, L_1, and C_1. A small gap is formed between the paired $\lambda/4$ resonators and the radiating patch, providing a capacitive coupling that can be modeled as a π-type network with three capacitances, C_g, C_{s1}, and C_{s2}. The T-shaped side-fed structure is equivalent to two $\lambda/4$ resonators, and the relevant parameters are about C_2 and L_2. The reactance introduced by this pin is taken into consideration and is reasonably modeled as a reactance L_{pin}.

The first step is to determine the sizes of the patch ($L_p \times W_p$) and $\lambda/4$ resonators ($L_r \times W_r$) according to the required central frequency f_0. The resonant frequencies of the radiating patch and $\lambda/4$ resonator are mainly dependent on their lengths, L_p and L_r, respectively. The value of L_p and L_r could be acquired based on the resonant frequency.

The second step is to investigate the working principle of high-order mode suppression. Due to the mismatching of these modes, undesired high-order radiations can be effectively suppressed by the proposed technique shown in Figure 3.20.

To verify the effectiveness of this technique, a prototype antenna is designed, investigated, and compared with traditional ones. Both are operated at 4.9 GHz and are fabricated on Rogers Duroid 5870 substrates (ε_r = 2.33 and h = 1.57 mm). Figure 3.22 also gives an S-parameter comparison between a traditional MPA and our proposed MPA. The traditional MPA generates a set of minimum poles in a wide-frequency range due to the matched higher order

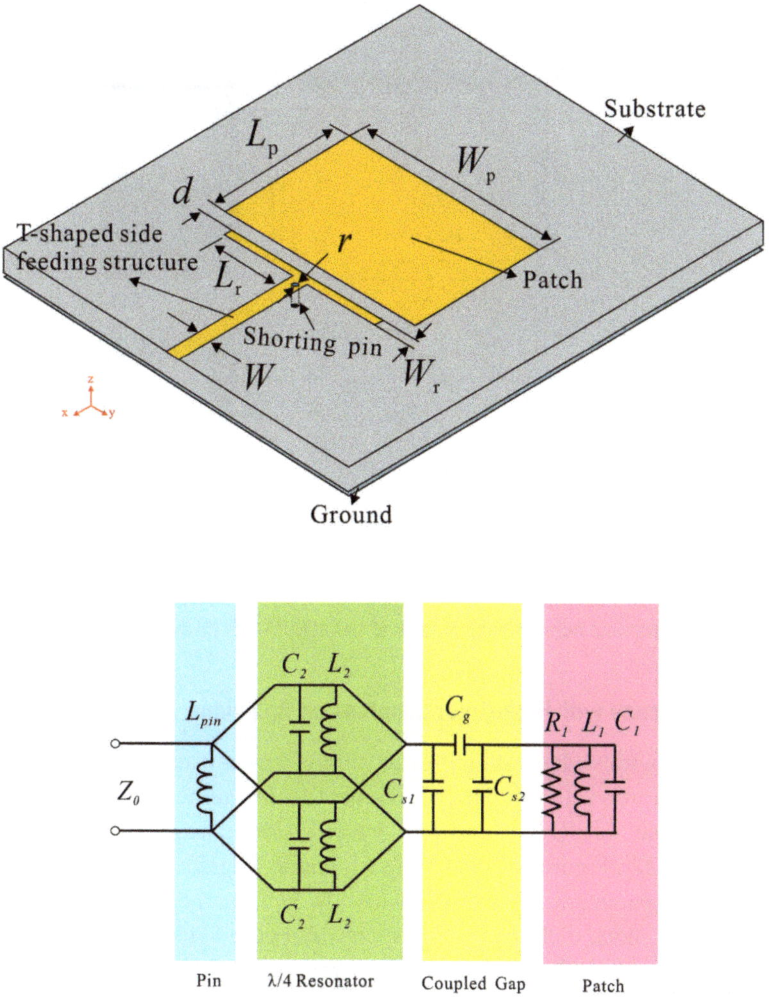

Figure 3.21 Configuration and equivalent circuit model of a low-profile multimode LP antenna with high-order suppression.

modes. Our proposed antenna with coupled feeding successfully suppresses most of the higher modes ranging from 4 to 13 GHz. The simulated and measured results agree with each other in the entire plotted range.

By using a pair of $\lambda/4$ resonators in this part, the bandwidth of the patch antenna is significantly enlarged and the harmonic radiations are effectively suppressed, while other advantages of the patch antenna, such as low cost, low profile, and easy integration, remain. In the analysis, an equivalent circuit model is proposed to analyze and design this proposed antenna. Our investigation shows that the bandwidth of this antenna can be widened by adjusting the gap between the patch and the $\lambda/4$ resonator.

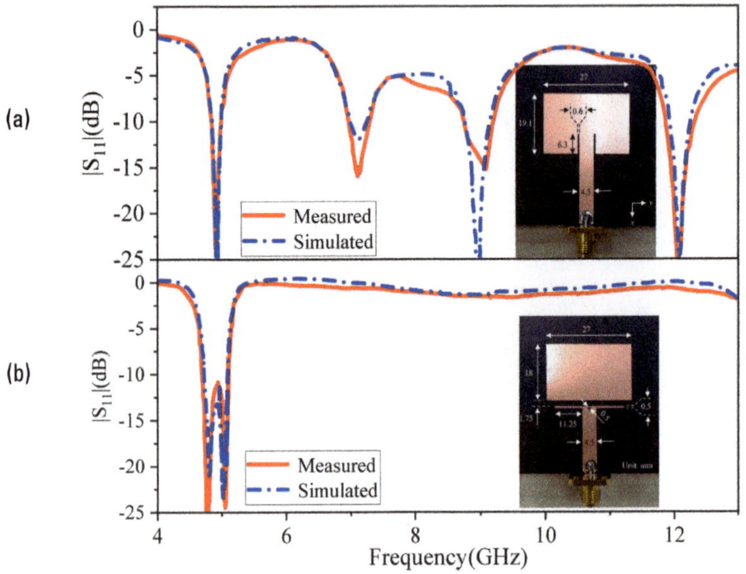

Figure 3.22 Harmonic comparisons of (a) traditional MPA and (b) our proposed MPA.

3.4.2 CP Antenna with High-Order Suppression Technique

CP antennas with wideband high-order suppression are also important in wireless communication. In this part, we discuss the design method of the triple-polarized antenna (LP, RHCP, and LHCP) with simultaneously wide high-order suppression [27]. Figure 3.23 plots the schematic diagram and configuration of the proposed patch antenna, which is mainly composed of the radiation patch, shorting pins, slots, dielectric substrate, feeding port 1, feeding port 2, feeding port 3, and ground plane. The specific parameters of the antenna are: $L = 55$ mm, $L_1 = 41.8$ mm, $L_2 = 32.65$ mm, $L_3 = 16.5$ mm, $L_4 = 25$ mm, $W = 130$ mm, $W_1 = 50$ mm, $W_2 = 3.2$ mm, $W_3 = 2.5$ mm, $S_1 = 24.5$ mm, $S_2 = 22$ mm, $S_3 = 14.2$ mm, $R_1 = 0.4$ mm, $D_1 = 23.4$ mm, $H = 4$ mm. The dielectric substrate has a relative permittivity of 2.65 and a thickness of 4 mm. The antenna has characteristics of low profile, small size, triple polarization, high isolation, high efficiency, and high-order suppression.

First, to achieve high-order suppression performance in a CP antenna, shorting pins need to be loaded on the edges of the radiation patch along the x-axis and y-axis as in Figure 3.23. These shorting pins could effectively suppress TM_{10}, TM_{01}, TM_{20}, TM_{02}, TM_{21}, TM_{12}, TM_{22}, TM_{30}, TM_{03}, TM_{32}, TM_{23}, and TM_{40} modes of a traditional rectangular patch antenna. Only three modes, such as mode 1, mode 2, and mode 3 of a traditional MPA could be finally excited, which are derived from TM_{11}, TM_{31}, and TM_{13} modes. Figure 3.24 plots the relative electric-field distributions of the shorted antenna under these

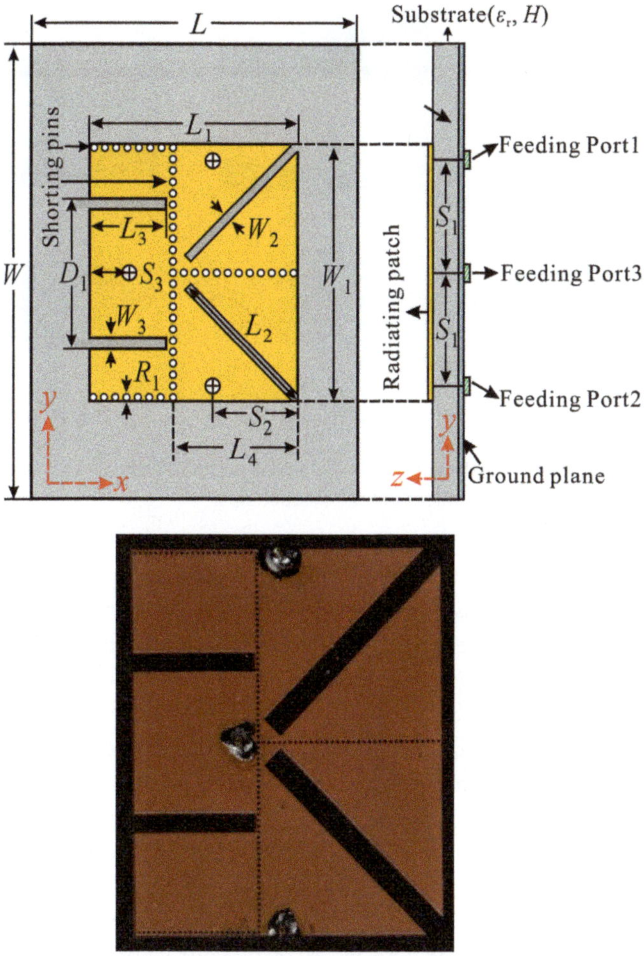

Figure 3.23 Configuration of our proposed low-profile triple-polarized MPA with simultaneous wideband high-order suppression.

three modes. The electric-field areas of mode 1, mode 2, and mode 3 are only a quarter of the infield areas of the primary TM_{11}, TM_{31}, and TM_{13} modes. Most important, the electric fields of mode 1 and mode 3 are orthogonal with each other, which can be used for CP generation by cutting the linear slot on the corner of the radiator.

To illustrate variation of these modes more clearly, Figure 3.25 gives the S-parameter comparison between a traditional patch antenna, the shorted antenna (antenna 1), and the shorted antenna with a linear slot (antenna 2). Over a bandwidth ranging from 1.5 to 7.1 GHz, multiple resonant modes can be simultaneously excited within a single MPA. It is a challenge to suppress some of the high-order modes as desired. For antenna 1, there are only three resonant

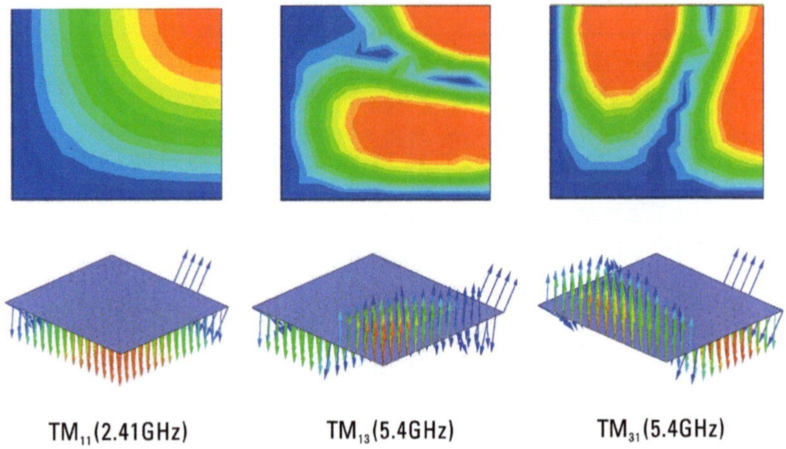

Figure 3.24 Electric-field distribution of the shorted antenna at two edges of the patch under mode 1, mode 2, and mode 3, under different frequencies.

modes in the frequency range of 1.5 to 7.1 GHz due to the mode suppression from these edge pins. The resonant frequencies of mode 2 and mode 3 are maintained almost the same, and the frequency ratio between mode 1 and mode 3 keeps a large ratio due to the different electric-field distribution as shown in Figure 3.24.

The electric fields of mode 1 and mode 3 are orthogonal with each other, and the slot is cut around the electric-field null of mode 3. The resonant frequency of mode 3 is significantly moved down, while keeping almost constant for that of mode 2. By cutting the slot for antenna 2, the frequency ratio between mode 1 and mode 3 is significantly reduced. Employment of the linear slot could make the input impedance of mode 2 under the mismatching condition, resulting in the antenna with the attractive out-of-band high-order mode-suppression characteristics in a wide-frequency band.

Apart from the realization high-order mode suppression for the compact CP antenna, it is also highly demanded to discuss the LP antenna with the desired high-order suppression performance. Figure 3.26 shows the configuration and *S*-parameter comparison between the traditional patch antenna, the shorted patch antenna (antenna 3), and the shorted antenna with linear slots (antenna 4). It shows that the traditional MPA has generated several resonant modes ranging from 1.5 to 7.1 GHz. But for the antenna 3, only mode 1 and mode 2 could be excited during the wideband in the range of 1.5 to 7.1 GHz due to the shorted pins around the core radiator edges.

Unfortunately, mode 2 is still resonated during this working band. As such, a pair of linear slots is cut on the radiating patch. The results show that the S-parameter characteristics of the antenna 4 with the loading of slot and shorting pin 3 only resonates in mode 1 in the range of 1.5 to 7.1 GHz. Hence, the

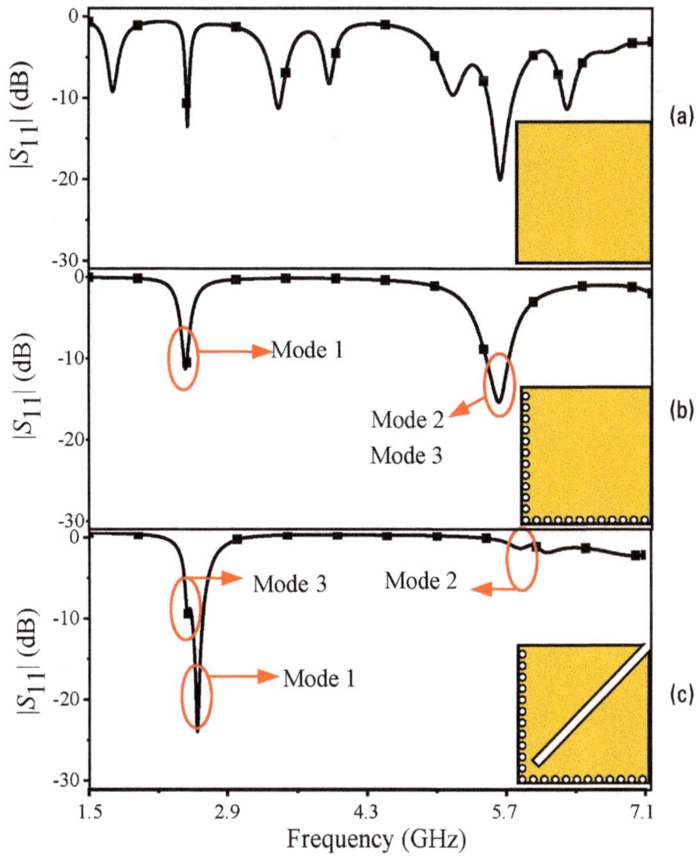

Figure 3.25 *S*-parameters comparison between a traditional microstrip patch antenna, antenna 1, and antenna 2: (a) traditional antenna, (b) antenna 1, and (c) antenna 2.

antenna has successfully gained the high-order suppression characteristics in the wide working band. In addition, both the CP antenna in Figure 3.25(c) and the LP antenna in Figure 3.26(c) have the shorted-edge scheme, it is only necessary to share the shorting wall of the abovementioned antennas to maintain high-isolation characteristics between multiple ports.

With these arrangements, high-order suppression performance is gained for both LP and CP antennas. The proposed triple-polarized antenna is designed, fabricated, and measured in Figure 3.23. A traditional triple-polarized antenna with three independent radiating patches is used for performance comparison. Due to the shared shorted edges, our proposed triple-polarized antenna is significantly smaller than a traditional triple-polarized antenna.

Figure 3.27 shows the S-parameter comparison between these two types of patch antennas. The traditional MPA has a large number of resonant modes in the upper band, but the proposed triple-polarized patch antenna only

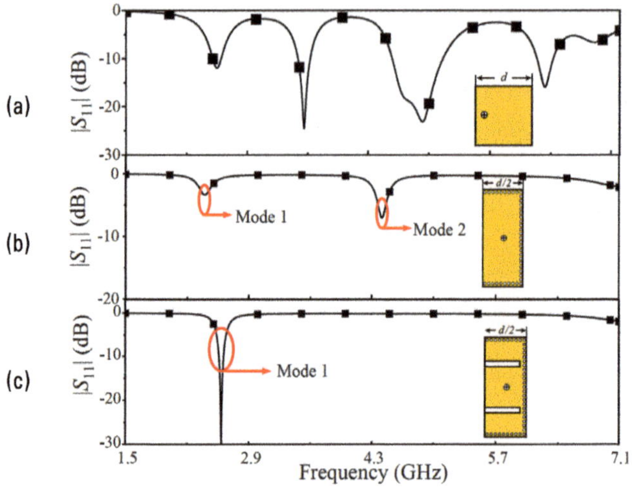

Figure 3.26 *S*-parameters comparison between the traditional MPA, antenna 3, and antenna 4: (a) traditional MPA, (b) antenna 3, and (c) antenna 4.

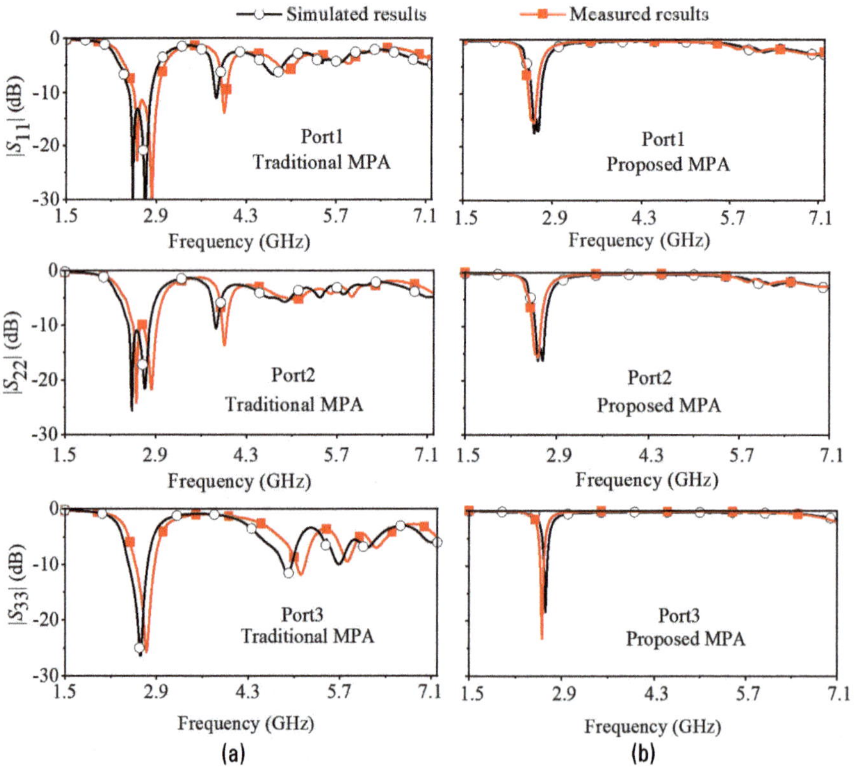

Figure 3.27 S-parameter comparison between two types of triple-polarized patch antennas: (a) traditional tripled-polarized microstrip patch antenna, and (b) proposed triple-polarized microstrip patch antenna.

maintains good matching characteristics near 2.65 GHz, achieving excellent out-of-band high-order suppression properties.

The shorting pins and slots are loaded to realize compactness and CP performance of the patch antenna, while the high-order suppression property is successfully gained for the antenna. High-order suppression of the LP antenna is gained by loading these pins and slots. With common shared shorting pins, these LP and CP antennas could be combined to generate triple-polarized radiation patterns under multiport feeding. The proposed multimode triple-polarized patch antenna maintains several attractive characteristics: low profile, triple polarization, small size, high-order suppression, and high isolation. Previous works have been proposed and developed using the multimode concept shown in Figure 3.28. In [28], the shorting pins and slots are inserted into the radiating patch to turn up the resonant frequencies of upper modes, resulting in 2nd and 3rd harmonic tuning as well as dual-frequency operation. In [29], the conventional rectangular slot antenna and double spur-line in the rectangular slot connected with ground plane are adopted to realize wideband harmonic suppression performance as desired.

3.5 Conclusions

Impedance bandwidth performance of antennas can be enhanced by using several techniques, such as stacked patches, L-shaped feeding probe, parasitic elements, array antennas, and so on. But compared to these mentioned methods, multimode resonance-inspired antennas with wide impedance bandwidth in this section have the advantages of compact size, wide impedance bandwidth, low profile, high efficiency, single layer, and single element. These proposed

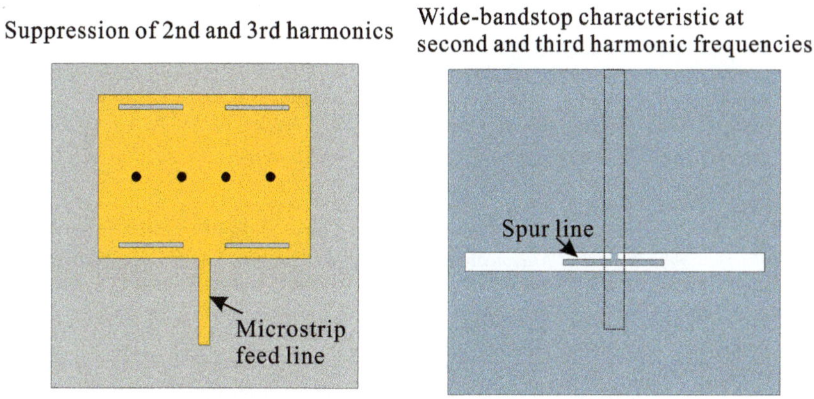

Figure 3.28 Harmonic suppression antennas under multimode resonance.

antennas have potential applications in modern communication systems, such as mobile phone devices, conformal systems, wearable technologies, and so on.

References

[1] Wu, Y., et al., "Broadband Circularly Polarized Patch Antenna with Parasitic Strips," *IEEE Antennas and Wireless Propagation Letters,* Vol. 14, November 2015, pp. 559–562.

[2] Deng, H., et al., "A Novel Feeding Network for Quadri-Polarization Antennas with Simultaneously Wide Impedance and Polarization Bandwidth," *IEEE Transactions on Antennas and Propagation,* Vol. 66, No. 4, April 2018, pp. 2107–2112.

[3] Lin, F. -H., and Z. -N. Chen, "Low-Profile Wideband Metasurface Antennas Using Characteristic Mode Analysis," *IEEE Transactions on Antennas and Propagation,* Vol. 65, No. 4, April 2017, pp. 1706–1713.

[4] Luk, K. -M., et al., "Broadband Microstrip Patch Antenna," *IET Electronics Letters,* Vol. 34, No. 15, July 1998, pp. 1442–1443.

[5] Wong, K. -L., and T. -W. Chiou, "Finite Ground Plane Effects on Broadband Dual Polarized Patch Antenna Properties," *IEEE Transactions on Antennas and Propagation,* Vol. 51, No. 4, April 2003, pp. 903–904.

[6] Zhu, L., S. Sun, and W. Menzel, "Ultra-Wideband (UWB) Bandpass Filters Using Multiple-Mode Resonator," *IEEE Microwave and Wireless Components Letters,* Vol. 15, No. 11, November 2005, pp. 796–798.

[7] Xiao, S. -Q., et al., "Bandwidth-Enhancing Ultralow-Profile Compact Patch Antenna," *IEEE Transactions on Antennas and Propagation,* Vol. 53, No. 11, November 2005, pp. 3443–3447.

[8] Liu, J., et al., "Design and Analysis of a Low-Profile and Broadband Microstrip Monopolar Patch Antenna," *IEEE Transactions on Antennas and Propagation,* Vol. 61, No. 1, January 2013, pp. 11–18.

[9] Lu, W. -J., and L. Zhu, "Wideband Stub-Loaded Slotline Antennas Under Multi-Mode Resonance Operation," *IEEE Transactions on Antennas and Propagation,* Vol. 63, No. 2, February 2015, pp. 818–823.

[10] Lu, W. -J., and L. Zhu, "Planar Dual-Mode Wideband Antenna Using Short-Circuited-Strips Loaded Slotline Radiator: Operation Principle, Design and Validation," *International Journal of RF and Microwave Computer-Aided Engineering,* Vol. 25, No. 7, September 2015, pp. 573–581.

[11] Lu, W. -J., et al., Wideband Dipole Antenna Using Multi-Mode Resonance Concept, *International Journal of Microwave and Wireless Technologies,* Vol. 9, No. 2, March 2017, pp. 365–371.

[12] Liu, N. -W., et al., "A Novel Differential-Fed Patch Antenna on Stepped-Impedance Resonator with Enhanced Bandwidth Under Dual-Resonance," *IEEE Transactions on Antennas and Propagation,* Vol. 64, No. 11, November 2016, pp. 4618–4625.

[13] Pues, H., and A. -V. -D. Capelle, "Accurate Transmission-Line Model for the Rectangular Microstrip Antenna," *IEE Proceedings H (Microwaves, Optics and Antennas)*, Vol. 131, No. 6, December 1984, pp. 334–340.

[14] Liu, N. -W., et al., "A Low-Profile Aperture-Coupled Microstrip Antenna with Enhanced Bandwidth Under Dual-Resonance," *IEEE Transactions on Antennas and Propagation*, Vol. 65, No. 3, March 2017, pp. 1055–1062.

[15] Liu, N. -W., et al., "Wideband Shorted Patch Antenna Under Radiation of Dual-Resonant Modes," *IEEE Transactions on Antennas and Propagation*, Vol. 65, No. 6, June 2017, pp. 2789–2796.

[16] Gao, S., et al., "Low-Profile Microstrip Patch Antenna with Improved-Bandwidth by Exciting Multi-Mode of the MPA and Network," *2021 International Applied Computational Electromagnetics Society (ACES-China) Symposium*, July 2021, pp. 1–2.

[17] Lu, W. -J., et al., "Design Approach to a Novel Dual-Mode Wideband Circular Sector Patch Antenna," *IEEE Transactions on Antennas and Propagation*, Vol. 65, No. 10, October 2017, pp. 4980–4990.

[18] Zhong, S. -S., and Y. -T. LO, "Single Element Rectangular Microstrip Antenna for Dual Frequency Operation," *IET Electronics Letters*, Vol. 19, No. 8, April 1983, pp. 298–300.

[19] Liu, J., and Q. Xue, "Broadband Long Rectangular Patch Antenna with High Gain and Vertical Polarization," *IEEE Transactions on Antennas and Propagation*, Vol. 61, No. 2, February 2013, pp. 539–546.

[20] Wang, B., and Y. -T. Lo, "Microstrip Antennas for Dual-Frequency Operation," *IEEE Transactions on Antennas and Propagation*, Vol. 32, No. 9, September 1984, pp. 938–943.

[21] Chair, R., et al., "Wideband Half U-slot Patch Antennas with Shorting Pin and Shorting Wall," *IEEE Antennas and Propagation Society Symposium*, Vol. 4, September 2004, pp. 4132–4135.

[22] Liu, N.-W., et al., "Low-Profile Triple-Band Microstrip Antenna Via Sharing a Single Multi-Mode Patch Resonator," *IET Microwaves, Antennas & Propagation*, Vol. 13, No. 10, August 2019, pp. 1580–1585.

[23] Liu, N. -W., et al., "Frequency-Ratio Reduction of a Low-Profile Dual-Band Dual-Circularly Polarized Patch Antenna Under Triple Resonance," *IEEE Antennas and Wireless Propagation Letters*, Vol. 19, No. 10, October 2020, pp. 1689–1693.

[24] Yang, K. -P., and K. -L. Wong, "Dual-Band Circularly-Polarized Square Microstrip Antenna," *IEEE Transactions on Antennas and Propagation*, Vol. 49, No. 3, March 2001, pp. 377–382.

[25] Kumar, K., S. Dwari, and M. -K. Mandal, "Dual-Band Dual-Sense Circularly Polarized Substrate Integrated Waveguide Antenna," *IEEE Antennas and Wireless Propagation Letters*, Vol. 17, No. 3, March 2018, pp. 521–524.

[26] Zhang, J. -D., et al., "A Compact Microstrip-Fed Patch Antenna with Enhanced Bandwidth and Harmonic Suppression," *IEEE Transactions on Antennas and Propagation*, Vol. 64, No. 12, December 2016, pp. 5030–5037.

[27] Liu, N. -W., et al., "Design Approach for Low-Profile Tripolarization Patch Antenna with Simultaneous Harmonic Suppression," *IEEE Transactions on Antennas and Propagation,* Vol. 70, No. 4, April 2022, pp. 2401–2410.

[28] Kwon, S., H. -K. Yoon, and Y. -J. Yoon, "Harmonic Tuning Antennas Using Slots and Short-Pins," *IEEE Antennas and Propagation Society International Symposium,* Vol. 1, July 2001, pp. 118–121.

[29] Kim, H., and Y. -J. Yoon, "Microstrip-Fed Slot Antennas with Suppressed Harmonics," *IEEE Transactions on Antennas and Propagation,*, Vol. 53, No. 9, September 2005, pp. 2809–2817.

4

Multimode Resonance-Inspired Planar Antennas with Isolation-Bandwidth Improvement

4.1 Introduction

In MIMO wireless systems, multiple-port antennas at the transmitters and receivers are often used to increase signal capacity and optimize data throughput [1]. In size-limited communication devices, MIMO antennas should be as integrated as possible and coexist with other antennas [2]. Unfortunately, the weak isolation of tight-layout antennas affects the overall performance of MIMO systems.

In this background, antenna bandwidth is not only focused on its impedance matching but on interelement isolation performance. The isolation bandwidth of planar antennas will be discussed and enhanced by using the even- and odd-order mode-resonance technique in Section 4.2, coupling-interface technique in Section 4.3, and mode-combination technique in Section 4.4.

4.2 Even- and Odd-Order Mode-Resonance Technique

A general mutual-conductance-eliminating principle is introduced in Section 4.2.1. It is then employed to yield decoupled designs of inverted-F antennas, and narrowband and wideband MPAs with enhanced port isolation in the following sections.

4.2.1 Working Principle

First, through establishing the equivalent-circuit models with lossy J/K inverters, we find that the two symmetric radiators will be decoupled when the odd and even modes have identical resonant frequencies ($f_{odd} = f_{even}$) and modal conductance ($G_{odd} = G_{even}$).

The traditional L/C loading method simply adjusts odd and even modes to the same frequency, reducing the mutual susceptance B_{12}. The mutual conductance G_{12} arises from the difference in modal conductance between odd and even modes. Characteristic mode analysis (CMA) is then employed to create three types of independent antenna pairs, showing that altering the shape of antennas and reconfiguring the current distribution can yield nearly equal modal conductance. A semianalytical design process can help achieve both high-isolation and strong-diversity performance.

For a pair of closely spaced antennas, there will be a strong interaction of electromagnetic fields between two radiators. In active near-field region, there are two coupling types through the fringing fields in proximity. One is the capacitive coupling through electric fields; the other is the inductive coupling through magnetic fields. The coupling coefficient k is a function of the dot product of their electromagnetic fields, that is, $\mathbf{E}_1 \cdot \mathbf{E}_2$ and $\mathbf{H}_1 \cdot \mathbf{H}_2$, if the far-field interaction is not considered.

Different from the reactive coupling between lossless resonators, mutual conductance exists between two radiators, which accounts for far-field interaction [3]. The mutual conductance G_{12} [4] is given by

$$G_{12} = \frac{1}{|V_0|^2} \mathrm{Re} \iint \mathbf{E}_1 \times \mathbf{H}_2^* \cdot ds \tag{4.1}$$

where $\mathbf{E}_1$ and $\mathbf{H}_2$ are the vectors of electric and magnetic fields radiated by antennas #1 and #2, respectively, while V_0 is the amplitude of excitation voltage.

There are both electric and magnetic couplings between two radiators. In electric coupling, a symmetric antenna pair can be equivalent to a two-port circuit network in Figure 4.1 with solid lines. The antenna is characterized by a parallel GLC resonator and the coupling is modeled as a lossy J inverter ($J = \omega C_m - jG_m$) where the mutual admittance comes from both mutual capacitance C_m and conductance G_m [3, 5]. The antenna will support both odd and even modes. According to the results in [6], because of an electric-coupling effect, the original characteristic frequency f_{odd} for the odd mode is lower than f_{even} for the even mode, as seen in (4.2).

$$f_{odd} = \frac{1}{2\pi\sqrt{L(C+C_m)}} < f_{even} = \frac{1}{2\pi\sqrt{L(C-C_m)}} \tag{4.2}$$

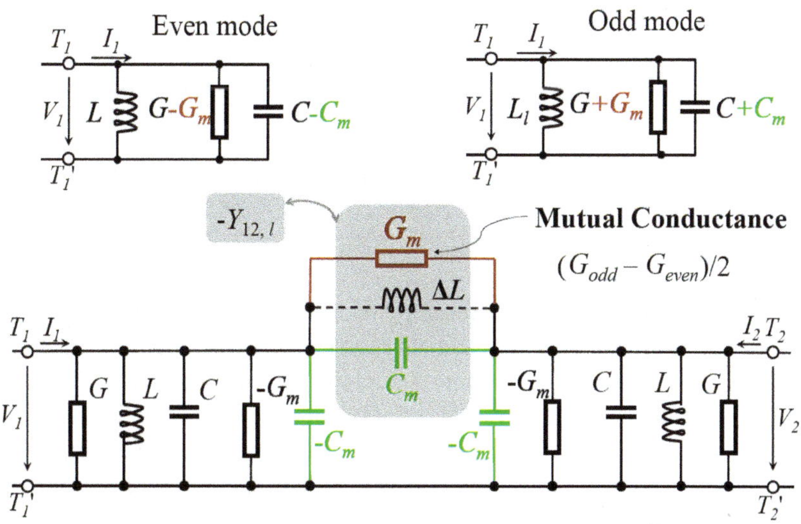

Figure 4.1 The equivalent-circuit model of an antenna pair coupled via a lossy J inverter and its decoupling technique with an external inductor. The simplified circuits for odd mode and even mode are also illustrated.

In this case, decoupling is to suppress the mutual admittance $Y_{12,l}$ to zero. To satisfy this requirement, a lumped inductor ΔL is loaded at the symmetric plane of the circuit model in Figure 4.1, which constructs a shunt resonator with the mutual capacitance C_m. At the resonant frequency f_0 of this resonator, the mutual susceptance $B_{12,l}$ is equal to zero and we have

$$-Y_{12,l} = -jB_{12,l} = j(\omega_0 C_m - \frac{1}{\omega_0 \Delta L}) = 0 \tag{4.3}$$

By applying the electric and magnetic walls to the symmetric plane [7], the equivalent-circuit models for the odd and even modes, respectively, can be obtained, as shown in Figure 4.1. The even-mode frequency $f_{even,l}$ is unchanged, whereas the odd-mode frequency $f_{odd,l}$ rises due to the loaded inductor ΔL.

$$f_{even.1} = 1/2\pi\sqrt{L(C - C_m)} \tag{4.4a}$$

$$f_{odd.1} = 1/2\pi\sqrt{L_l(C + C_m)} \tag{4.4b}$$

By matching the isolation frequency f_0 to $f_{even,l}$, the required value of the loaded inductance ΔL can be derived as (4.5b). Simultaneously, $f_{odd,l}$ will be tuned to $f_{even,l}$

$$1/L_l = 1/L + 2/\Delta L \tag{4.5a}$$

$$\Delta L = L\left(C/C_m - 1\right) \tag{4.5b}$$

$$f_0 = f_{even,l} = f_{odd,l}. \tag{4.5c}$$

In a similar manner, the magnetic coupling can be eliminated by adding an external lumped capacitor, and the antenna becomes decoupled with the frequency $f_{even,l}$ tuned to $f_{odd,l}$. Note that dominant coupling type is more accurately determined by examining the relationship between resonant frequencies rather than field distribution [6]. The operational principle of the lumped-loading methods involves tuning the odd and even modes resonating to be equal and reducing the imaginary part of Y_{12} or Z_{12}.

Then, the mutual conductance must vanish for decoupling. According to symmetric circuit analysis [7], the modal conductance of the odd and even modes differs, because

$$G_{odd} = G + G_m \tag{4.6a}$$

$$G_{even} = G - G_m \tag{4.6b}$$

According to (4.6) and Figure 4.1, the mutual conductance between the two antennas can be derived as

$$-G_{12} = G_m = \left(G_{odd} - G_{even}\right)/2. \tag{4.7}$$

When $G_{odd} = G_{even}$, mutual conductance should vanish.

In a general case, for a closely spaced antenna pair, the mutual coupling is contributed by the mutual conductance G_m, mutual capacitance C_m, and mutual inductance L_m.

$$-Y_{12} = G_m + j\left(\omega C_m - 1/\grave{u} L_m\right) = 0 \tag{4.8}$$

Let $\mathrm{Im}(Y_{12}) = 0$, the resonant frequencies of the odd and even modes should be reallocated to be equal,

$$f_{odd} = f_{even} = f_0 \tag{4.9a}$$

To further reduce mutual conductance, the modal conductance of odd and even modes should be equal,

$$G_{odd} = G_{even} = G \tag{4.9b}$$

If both conditions in (4.9a) and (4.9b) are met, decent isolation will be produced at resonant frequency. To realize wideband isolation, quality factors should also be equal

$$Q_{odd} = Q_{even} = 2\pi f_0 GC \tag{4.9c}$$

where C is the capacitance of odd and even modes.

Based on the above theoretical discussion, we will use this method to improve the isolation bandwidth of the inverted-F antennas and patch antennas in the following.

4.2.2 Isolation of Two Inverted-F Antennas

In this section, we apply the proposed method to a pair of inverted-F antennas (IFAs), which are decoupled without lumped elements, and the decoupling is achieved only by shape modification. The geometries of the compared antennas in [8] printed on the infinite 0.508-mm-thick 4003C substrates are shown in Figure 4.2. In case 1, the two monopoles are straight in shape and parallel to each other. In cases 2 and 3, the two monopoles are bent with three segments parallel to the ground edges, and their vertical clearance height H_v is set differently. In these three cases, the resonant frequencies of the even and odd modes for each antenna have been adjusted to be equal, as indicated by results of modal conductance in Figure 4.2.

The reflection coefficient and port-to-port isolation in the three cases are shown in Figure 4.3. Across the working band, the isolation levels in cases 1, 2, and 3 are 3.3, 21.8, and 48.9 dB, respectively. It demonstrates that the suppressed mutual susceptance via equal f_{odd} and f_{even} does not ensure high port-to-port isolation between the two radiators. This is because the mutual conductance in cases 1 and 2 is still too large.

To verify our assumption, we look back to the modal conductance of three cases in Figure 4.2. The difference of modal conductance, ΔG, is as large as 194 mS in case 1, which will contribute to the largest mutual conductance and the weakest isolation. In cases 2 and 3, ΔG is reduced to 6.3 and 0.1 mS, respectively, and the isolation is gradually increased. In case 3, the isolation level is even improved to above 45 dB when ΔG is reduced to zero, as shown in Figure 4.3 and case 3 in Figure 4.2.

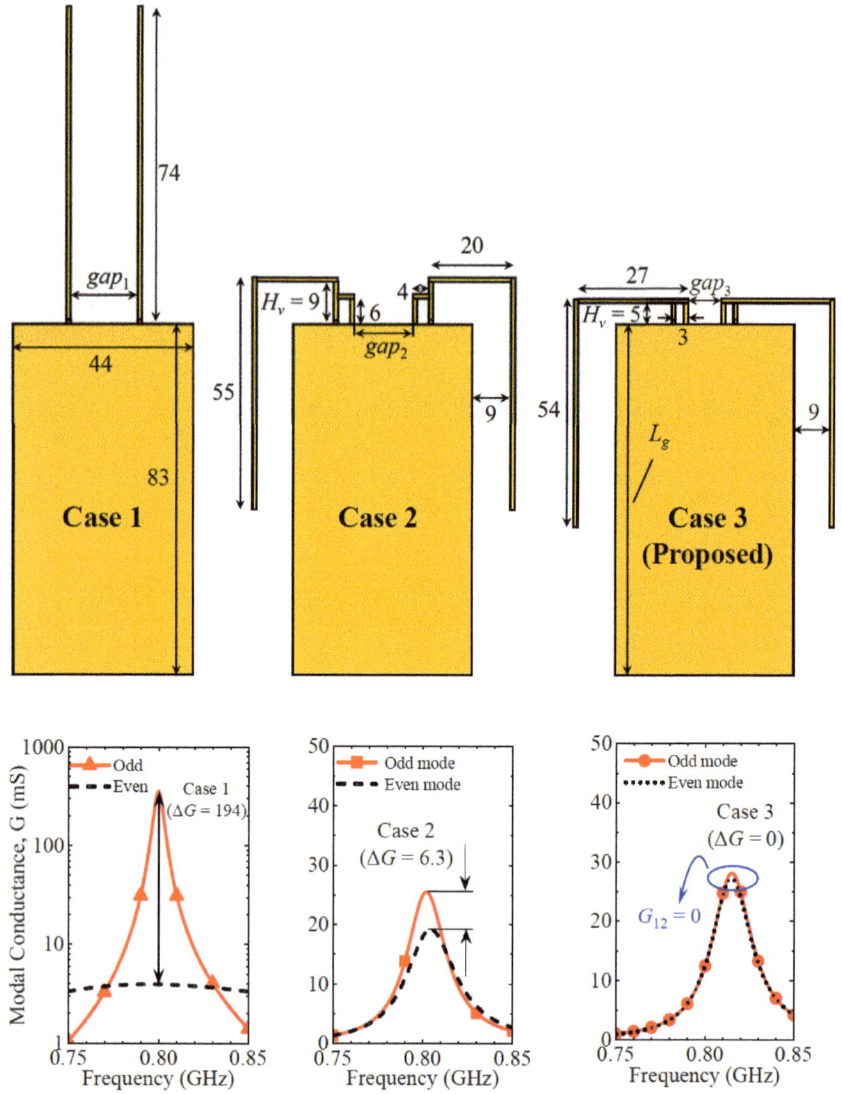

Figure 4.2 Evolution procedure and their modal conductance analysis of decoupled monopole antenna pair on a common ground plane. The infinite substrate is hidden, ($gap_1 = 16$, $gap_2 = 14$, $gap_3 = 8$). Units in the figure: mm.

The working mechanism of the varied modal conductance can be interpreted by the characteristic currents in Figure 4.3. In case 1, the in-phase currents of the even mode on the two monopoles contribute to enhanced radiation whereas the opposite currents of the odd mode result in suppressed radiation. The quality factors and modal conductance of the two modes are distinct, with $G_{even} < G_{odd}$ and $Q_{even} < Q_{odd}$. In case 3, with bent monopoles, each mode

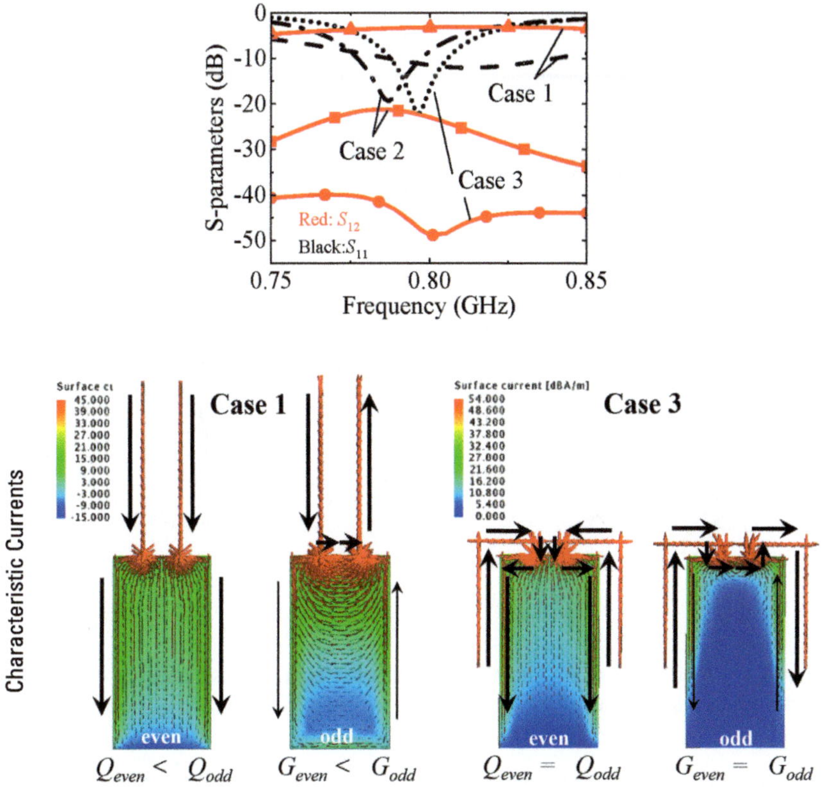

Figure 4.3 Simulated S-parameters of the antenna in Figure 4.2 and characteristic currents of case 1 and case 3.

consists of in-phase and out-of-phase current components. Even-mode radiation is partially suppressed while odd-mode radiation is partially enhanced in this way. The quality factors and modal conductance of the two modes become equal at last, and the two radiators are well decoupled.

According to the current distribution in Figure 4.3, the radiator spacing and ground size can manipulate the two modes and affect the isolation.

To validate the anticipated performance of the proposed self-cancellation antenna pair, an antenna prototype using lossy material was designed and manufactured, as shown in Figure 4.4. The antenna pair is printed on 0.508-mm-thick 4003C substrate ($\varepsilon_r = 3.55$, tan$\sigma = 0.003$) to facilitate fabrication via the PCB process. In this case, simulation is conducted using FEM, and the antenna size is slightly adjusted compared to that in case 3 in Figure 4.2 due to the consideration of additional loss. The antenna is fed by two 50Ω coaxial cables with two $\lambda/4$-open-circuit RF baluns [9] to mitigate radiation and coupling from the cables.

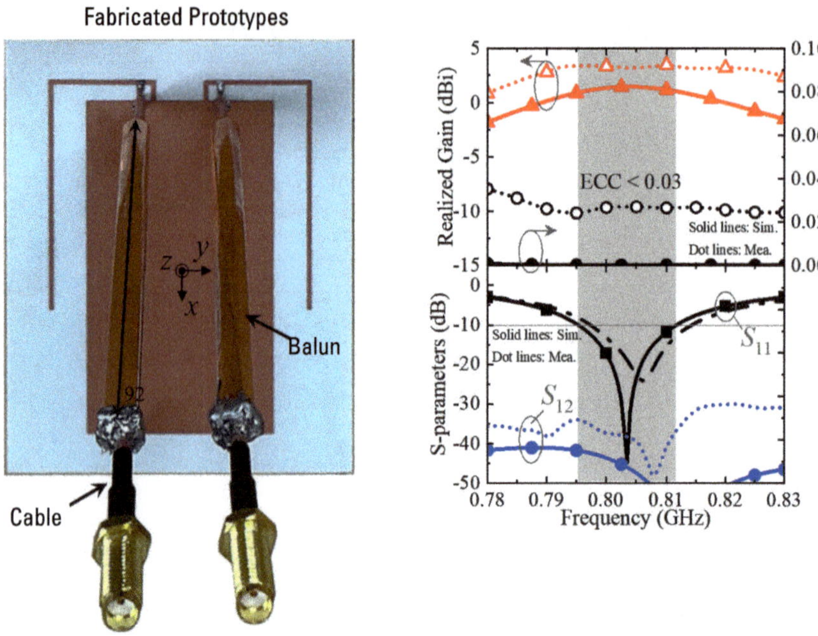

Figure 4.4 Photograph of the fabricated antenna prototype in case 3 and its simulated and measured results. Units: mm.

The simulated and measured S-parameters are also presented in Figure 4.4. The simulated 10-dB return-loss bandwidth ranges from 0.795 to 0.812 GHz, while the measured result spans from 0.797 to 0.814 GHz. Both demonstrate exceptional isolation, exceeding 40 dB. Additionally, as shown in Figure 4.4, the maximum realized gains in simulation and measurement are 1.5 and 3.4 dBi, respectively. Both the simulated and measured ECCs are below 0.03, indicating good diversity performance of the antenna.

4.2.3 Isolation of Two Narrowband Patch Antennas

This technique is also applicable to the decoupling of patch antenna pairs. As shown in Figure 4.5, three decoupling schemes of patch antenna pairs are compared under different configurations. In reference design 1, two square patches are tightly coupled to each other, and a lumped inductor is loaded at the central gap to suppress the mutual susceptance. In the proposed design 2, the original square patch antennas are modified into a parallelogram by adjusting the angle θ so that the modal conductance can be flexibly manipulated. In this section, all patch antennas to be modeled on an infinite ground plane are built on the 0.508-mm-thick 4003C substrates ($\varepsilon_r = 3.55$, tan$\sigma = 0.003$), and the two substrates are suspended with a 3-mm-thick air layer. CMA is conducted in the simulation with FEKO, and the metal and substrate are set lossless [10].

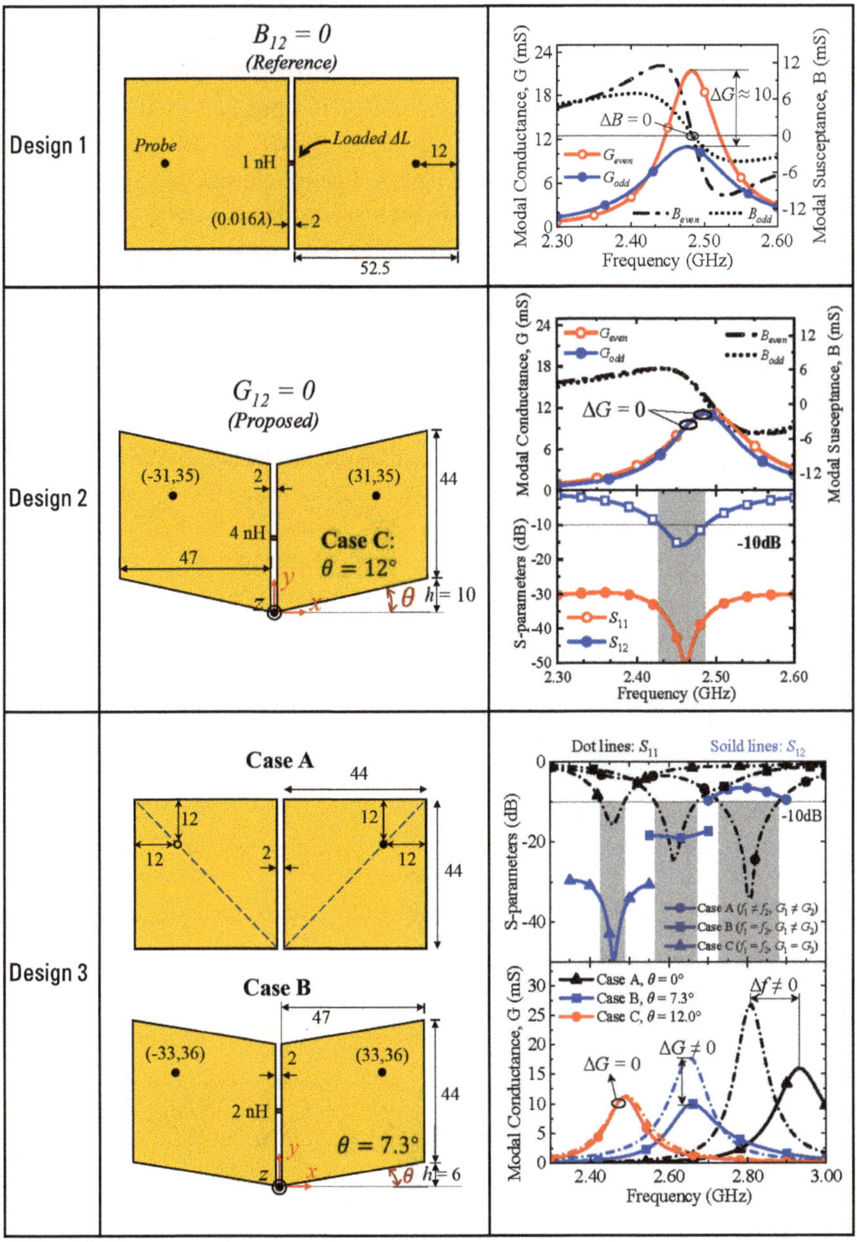

Figure 4.5 Three decoupling schemes of the patch antenna pair and their S-parameter and modal analysis. Units: mm.

In simulation, only the left port is fed by a voltage source to excite two modes simultaneously.

A pair of coupled square patch antennas without lumped loading are investigated with CMA simulation based on the antenna model in Figure 4.5 design 1 by removing the loaded inductor. The current distributions of excited odd and even modes are shown in Figure 4.6. The characteristic frequencies of odd and even modes show a notable difference ($f_{odd} < f_{even}$) caused by the strong electric coupling between the two radiators. The modal significance (MS) also manifests that the even mode has a higher quality factor, due to the stronger cancellation effect of its on-patch current. For these reasons, the isolation level of the antenna pair is only 4.7 dB.

Next, the study investigates the decoupling effect of the traditional loading method using the antenna in design 1, and its characteristic currents are depicted in Figure 4.6. The loaded inductor significantly disrupts the currents of the odd mode but has no impact on the even mode. Due to the inductive loading, the resonant frequency of the odd mode can be adjusted to match that of the even mode, which means the mutual susceptance should tend to vanish at the resonance. The isolation level improves from 4.7 to 14.5 dB. Specifically, the results in Figure 4.6 show that the mutual susceptance B_{12} is eliminated at 2.487 GHz, confirming the analysis and conclusion in Section 4.2.1.

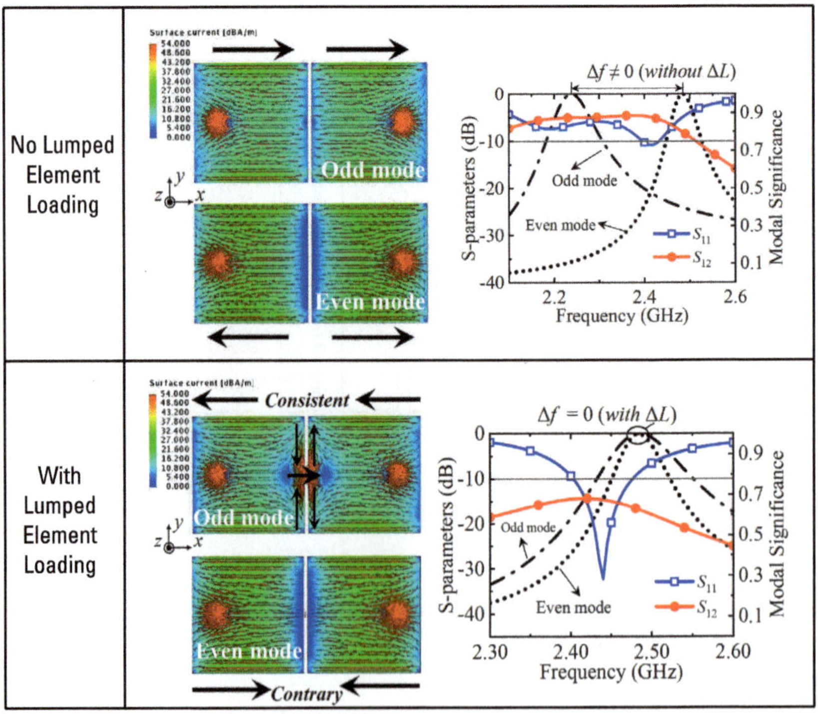

Figure 4.6 Simulated results of the coupled patch antenna pair with and without loading of lumped element in design 1 and design 2.

However, the 14.5-dB isolation is not sufficient, indicating that mutual coupling is not completely suppressed. This is primarily due to the high mutual conductance G_{12} in this case, which is 10.3 mS at the resonant frequency. Furthermore, as shown in the results of design 1 in Figure 4.5, the modal conductance of the two modes is evidently different, with a difference of ΔG also being 10.3 mS. This aligns well with the theoretical conclusion in (4.7) if the coefficient 1/2 for two-port excitation is replaced by 1 for one-port excitation. To fully achieve the desired isolation, the modal conductance of the odd and even modes should be equal.

The different modal conductance is caused by distinct modal current distributions. For the odd mode, both G_{odd} and Q_{odd} are lower, with the current on patches being in phase, making it easier to radiate. In contrast, the on-patch current of the even mode is out of phase, and the cancellation effect makes it less radiative. Consequently, G_{even} and Q_{even} are higher.

Modal conductance governs the radiation ability of the operating modes, so the primary approach to equalize G_{odd} and G_{even} is to reshape the current distribution of the two modes. To achieve this, the antenna pair in design 2 of Figure 4.5 is used, and the design follows two steps. Detailed analysis is also provided in our reported works in [8].

First, with the patch transformed into a parallelogram, the current vectors on each patch can be regarded as the superposition of TM_{01} and TM_{10} modes (i.e., $J_{01}^{y} \pm J_{10}^{x}$). The y-direction current components J_{01}^{y} of the odd mode are in opposite directions on two patches, causing certain cancellation effect of radiation. Similarly, the x-direction components J_{10}^{x} of the even mode are opposite and have similar effects.

Second, by changing the angle θ, the proportion of J_{01}^{y} and J_{10}^{x} will be rescaled and the conductance of the two modes will vary with different tendencies. For parallelogram patches in design 2, by increasing the y-direction current proportion of J_{01}^{y} via increased θ, the odd mode can be less radiative, and the even mode becomes more radiative. In this way, both quality factors Q_{odd} and Q_{even} and modal conductance G_{odd} and G_{even} of two modes are equal by changing the angle θ of the patch.

Comparing the modal conductance results of design 1 and 2 in Figure 4.5, the modal conductance G_{even} at resonance decreases from 21 to 11 mS, while G_{odd} increases from 10 to 11 mS. The maximum isolation level of the S-parameters of the antenna exceeds 40 dB, representing an enhancement of at least 25 dB compared to the design 1 antenna. The improved isolation is attributed to the reduced mutual conductance G_{12} via the modified antenna shape.

The isolation in this design does not stem from orthogonal polarization. To verify this, the proposed antenna is compared with the orthogonally fed square patch antenna pair in design 3, as shown in Figure 4.5. Although the two patches of the reference antenna in case A are excited with orthogonal modes, their isolation level is only 7 dB when they are closely spaced. This is due to the considerable difference in characteristic frequencies and modal conductance of the two modes, as shown in the analysis in Figure 4.5. Subsequently, the angle θ of the patch is increased by changing the variable h, and the loaded inductance is adjusted to maintain $f_{odd} = f_{even}$ in case B and case C. The corresponding modal conductance and S-parameters under different values of θ are also illustrated, respectively. The modal conductance difference ΔG at the characteristic frequencies is reduced from 6 to 0 mS when θ is increased from 7.3° to 12.0°, respectively. As a result, the isolation level in the band gradually improves from 18 to 34 dB. This means the proposed method is more effective than the orthogonal feeding scheme.

To validate the feasibility of the proposed decoupling method, a single-band antenna prototype was designed, fabricated, and measured. A photograph of the fabricated antennas is shown in Figure 4.7. The patches and ground are printed on 0.508-mm-thick 4003C substrates, and the two substrates are suspended with an air layer of 3-mm thickness.

Slightly different from the lossless material in CMA simulation, the real antennas have spurious dielectric loss and conduction loss. Thc simulation in this case is carried out with the finite element method (FEM), and its performance is tuned to determine the final dimensions of the antenna by compensating for the effect of additional loss. The lumped inductor is substituted by a connecting metal strip, because commercial lumped inductors are available only at specific values.

The simulated and measured results of the single-band-decoupled antenna pair are shown in Figure 4.7. As depicted, $|S_{11}|$ is below −10 dB in the frequency

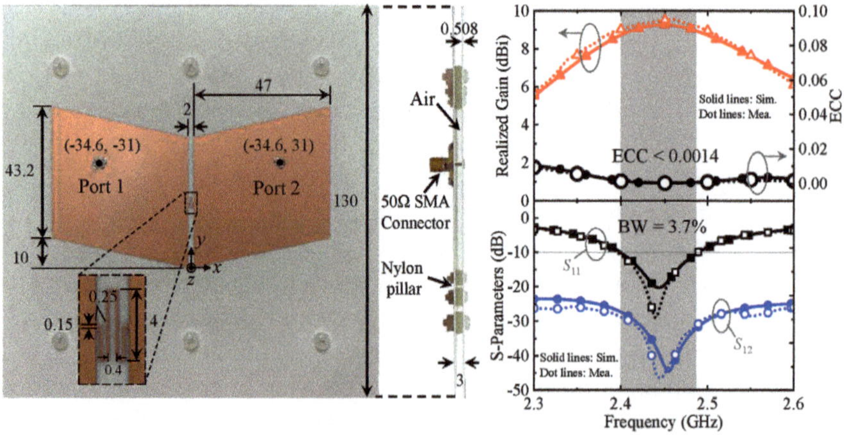

Figure 4.7 Photograph of the fabricated narrowband-decoupled patch antenna pairs. The simulated and measured S-parameters, broadside realized gain, and ECC of the narrowband-decoupled patch antenna pair are compared.

range from 2.40 to 2.49 GHz, and the in-band isolation level is over 27.4 dB. The simulated and measured broadside gains are 8.7 to 9.3 dBi and 9.0 to 9.7 dBi, respectively. Additionally, the ECCs of the antenna are calculated with far fields [11], and the values are low—less than 0.0014—demonstrating good diversity performance.

In conclusion, lumped loading can only suppress mutual susceptance; decoupling cannot be ultimate unless the modal conductance of the two modes differs little. Modified antenna shape is effective in manipulating modal conductance and improving isolation.

4.2.4 Isolation of Two Wideband Patch Antennas

The proposed decoupling technology can also be applied to the design of wideband patch antennas. The geometry of the proposed wideband antenna pair is depicted in Figure 4.8, where the aspect ratio of the parallelogram patch is increased so that high-order modes are all moved to low frequency. In characteristic mode-currents analysis, two pairs of odd and even modes are produced closely. The current distributions and radiation patterns of the two pairs of modes are also shown in Figure 4.8. The radiation patterns of modes 1 and 2 are complementary, while the radiation patterns of modes 3 and 4 are orthogonal. As such, they can be regarded as good candidates for MIMO antenna design.

Two key steps help achieve wideband isolation. First, the value and location of the loaded inductor are carefully chosen so that each pair of odd and even modes has the same characteristic frequencies ($f_{odd} = f_{even}$), as shown in Figure 4.9. This effectively reduces the mutual susceptance B_{12} to zero. Second, the modal conductance of each pair of modes is adjusted to be as equal as possible by altering the angle of each patch, further diminishing the mutual conductance G_{12}. It is explained in [8] that the change of the patch's angle affects the radiation/cancellation effect of the *x*- and *y*-components of currents on the patch under even and odd modes, which finally determines their quality factors. Therefore, the dimensional parameters of the antenna are optimized according to the change of current distribution and quality factors of even and odd modes.

Figure 4.9 displays the results of the modal conductance, showing that the modal conductance for each pair of odd and even modes has become nearly identical. As a result of the equal characteristic frequencies and modal conductance, this modified patch antenna achieves substantial wideband isolation. The 10-dB return-loss bandwidth spans from 3.49 to 3.89 GHz (10.8%), with in-band isolation ranging from 18 to 34 dB. However, the isolation of the wideband patch antenna pair is less effective at the lower band, as the modal conductance of the even and odd modes at the lower band is not strictly equal.

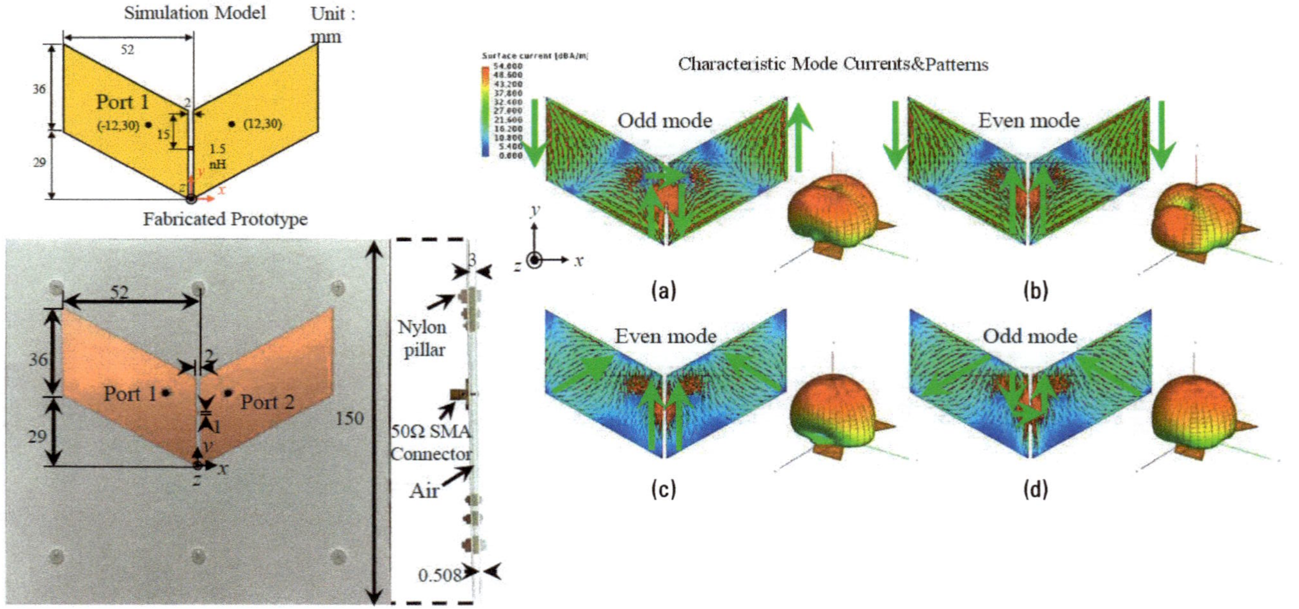

Figure 4.8 Wideband-decoupled patch antenna pair based on two pairs of odd and even modes. Currents distribution and radiation patterns of the modes are compared. Units: mm.

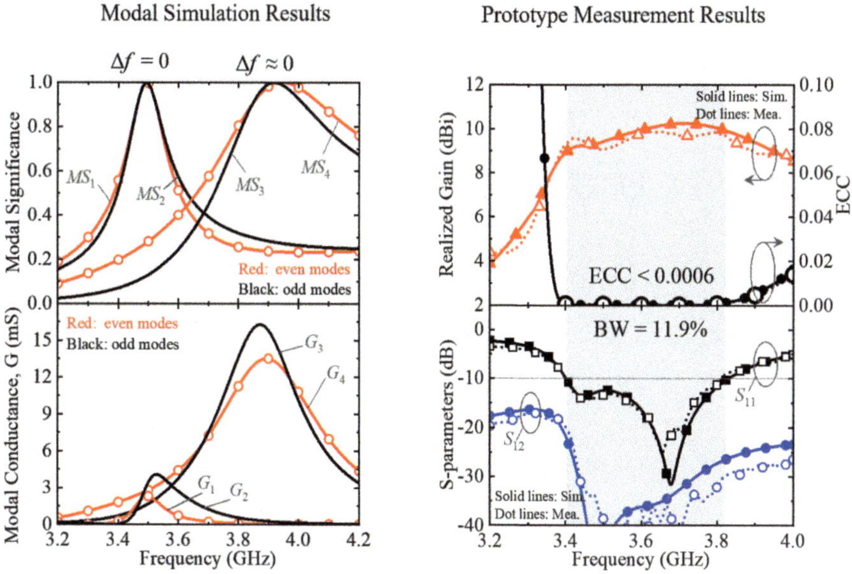

Figure 4.9 Modal simulation results of the proposed wideband patch antenna pair, and the measurement results of the fabricated prototype.

To validate the feasibility of the proposed decoupling method, an antenna prototype operating in the wideband was designed, fabricated, and measured. A photograph of the fabricated antennas is shown in Figure 4.8.

The simulated and measured results of the wideband-decoupled antenna pair are presented in Figure 4.9. Two reflection zeros have broadened the fractional bandwidth (FBW) of the 10-dB return loss to 11.9% (3.40 to 3.83 GHz). The simulated isolation level ranges from 21.8 to 53.2 dB across the operating bandwidth; the measured one ranges from 20.0 to 50.6 dB. The maximum realized gains in simulation and measurement are 10.2 and 9.9 dBi, respectively. Notably, the ECC of the antenna is maintained below 0.0006 in the operating band, demonstrating good diversity performance.

Unlike the use of odd- and even-mode decoupling, the mode-cancellation method considers decoupling from the perspective of equal scattering parameters of the CM and DM, as shown in Figure 4.10. In [12], it is theoretically demonstrated that if the CM and DM reflections are equal, the mutual coupling effect between two separate antennas can be eliminated. In [13], a self-decoupled antenna pair based on mode cancellation is proposed. By adjusting the distance D_0 between the two ports of the antenna pair, an isolation of better than 10.5 dB is achieved across the 5G N77 band. In [14], by inserting a lumped inductor between the closely spaced microstrip antennas, the CM and DM reflections are adjusted to be similar and an isolation is improved from 5 dB to better than 15.4 dB.

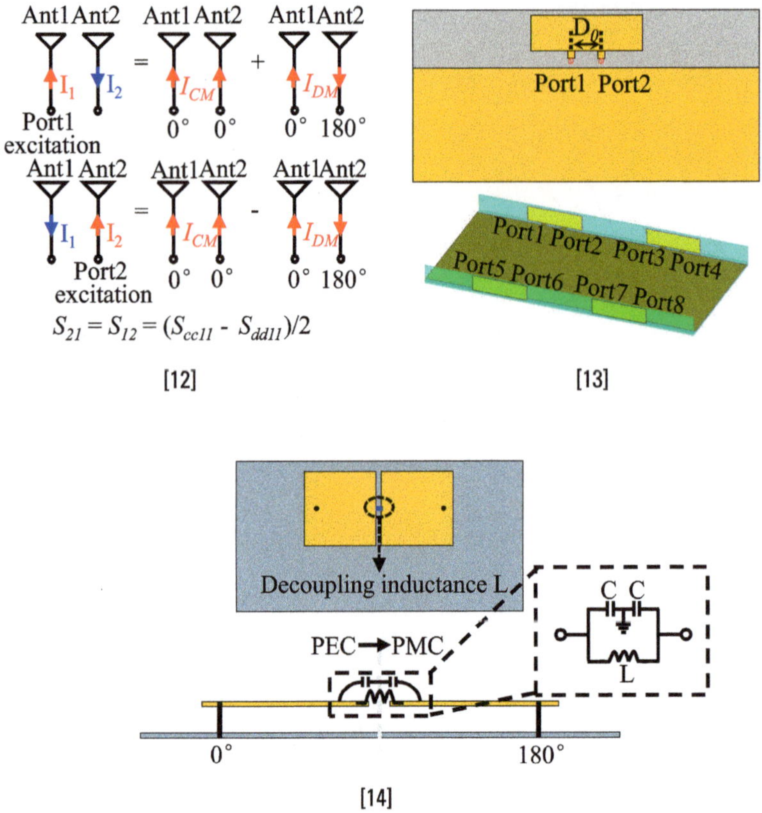

Figure 4.10 Illustration of common and differential modes cancellation decoupling.

4.3 Coupling-Interface Technique

The coupling-interface technique is another effective approach for isolation improvement, which consists of the coupling interface from via-loaded microstrip stripline in Section 4.3.1, the coupling interface from open-ended microstrip stripline in Section 4.3.2, and the coupling interface from short-circuited microstrip stripline in Section 4.3.3.

4.3.1 Coupling Interface from Via-Loaded Microstrip Stripline

Figure 4.11 plots the detailed structure and fabrication of the proposed dual-port antenna. The antenna consists of two rectangular patches with distance d_1, a half-wavelength transmission line, a shorting via centrally ground plane, and Rogers/Duroid 5880 dielectric substrate. The detail parameters of the antenna in [15] are: $L = 80$ mm, $L_p = 30$ mm, $L_f = 4.2$ mm, $L_m = 31$ mm, $W = 120$ mm,

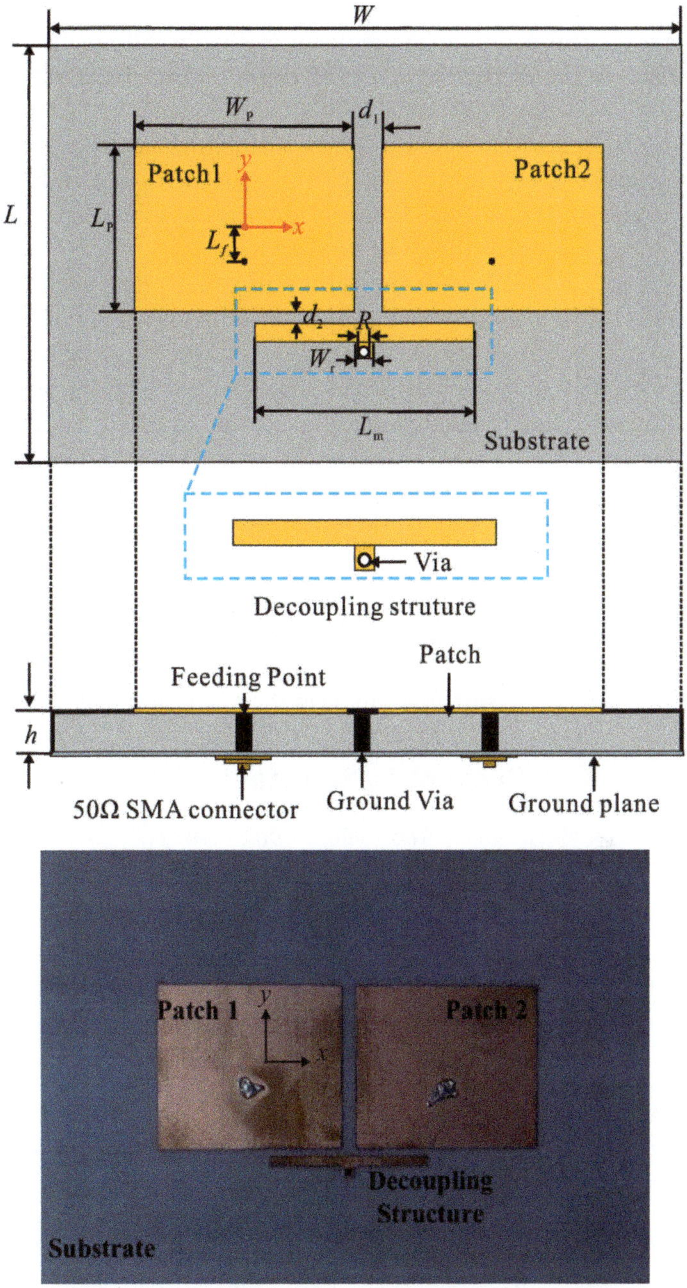

Figure 4.11 Geometry and fabrication of two-element patch antennas with the proposed decoupling structure.

W_p = 35.4 mm, W_m = 2 mm, W_r = 2 mm, d_1 = 2.6 mm, d_2 = 1.2 mm, R = 1.2 mm, h = 2 mm.

The half-wavelength transmission line in Figure 4.11 can provide a 180°-phase transformation at the resonant frequency. Meanwhile, the interference signal in the decoupling path will be transmitted from patch 1 to the transmission line with half-wavelength through the coupling interface, then coupled to patch 2. Additionally, these two couplings can be replaced by using the capacitive coupling with 90°-phase lag. A total phase shift of 360° is realized for the antenna under this scheme. A detailed analysis process with its working principle is discussed in the following.

Figure 4.12 has built the equivalent circuit of the proposed array antenna in Figure 4.11. The potentials at two open-ended transmission lines show the maximum distribution at its resonant frequency, while for the center of the microstrip stripline, it is an electric-field node accompanied by maximum current distribution. In this design, a small microstrip line with a shorted pin is used in parallel with the intermediate region of the transmission line. An inductance near the resonant frequency of the antenna is formed up. This additional inductance leads to a 90°-phase delay, so the signal from patch 1 to patch 2 through the decoupling scheme will have a total phase shift of 270° from a 180°-phase shift derived from two capacitive couplings, a 180°-phase shift derived from the half-wave transmission line, and a 90°-phase delay derived from the introduced inductor. The signal transmitted from patch 1 to patch 2 through the decoupling structure produces the phase shift near 270°. For the

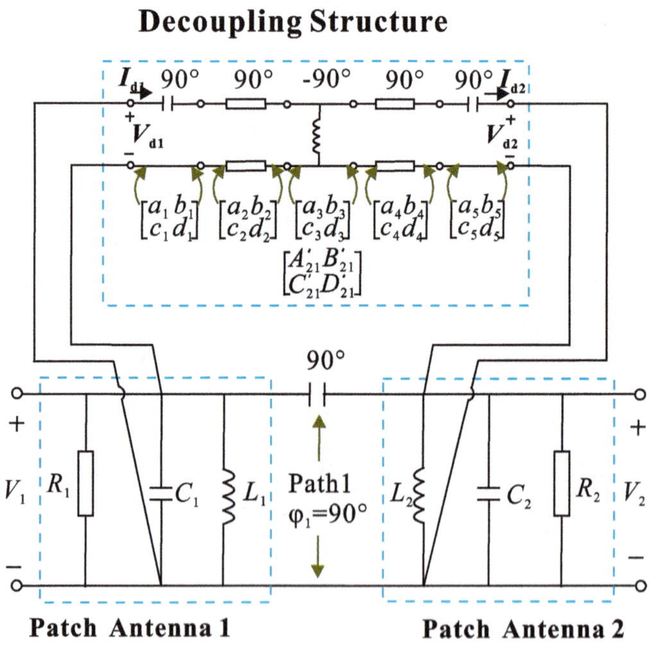

Figure 4.12 Equivalent circuit of two-element patch array with its decoupling structure.

decoupling structure, input and output voltages at the interface of patch 1 and patch 2 are set as V_{d1} and V_{d2}, respectively. By controlling these key parameters, the isolation performance of the antenna under two ports could be enhanced during the working band, which has not conducted the parameter studied in the following.

To validate the predicted isolation performance, the two-port MIMO antenna in Figure 4.11 was designed, simulated, and measured, which is operated around 3.16 GHz. Figure 4.13 gives the simulated and measured S-parameters with and without the via-loaded microstrip stripline. The results prove that the designed antenna has successfully gained the isolation improvement of about 11 dB by utilizing the decoupling structure.

All in all, the coupling interface technique based on via-loaded microstrip stripline has been demonstrated as a useful method for isolation improvement of two closely spaced patch antennas.

4.3.2 Coupling Interface from Open-Ended Microstrip Stripline

In this part, the open-ended microstrip stripline is verified as an alternatively method for isolation improvement. Figure 4.14 depicts the detail configuration of the proposed antenna, which consists of two identical rectangular patches and identical side-by-side half-wavelength microstrip lines. The detail parameters of the antenna in [16] are listed as follows: $W = 30$ mm, $L = 21.5$ mm, H

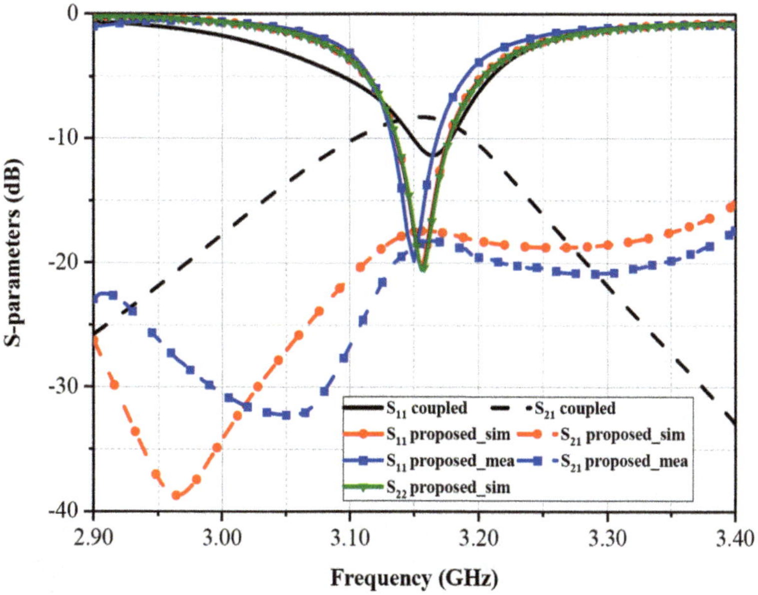

Figure 4.13 Simulated and measured S-parameters of the two-port MIMO patch antenna.

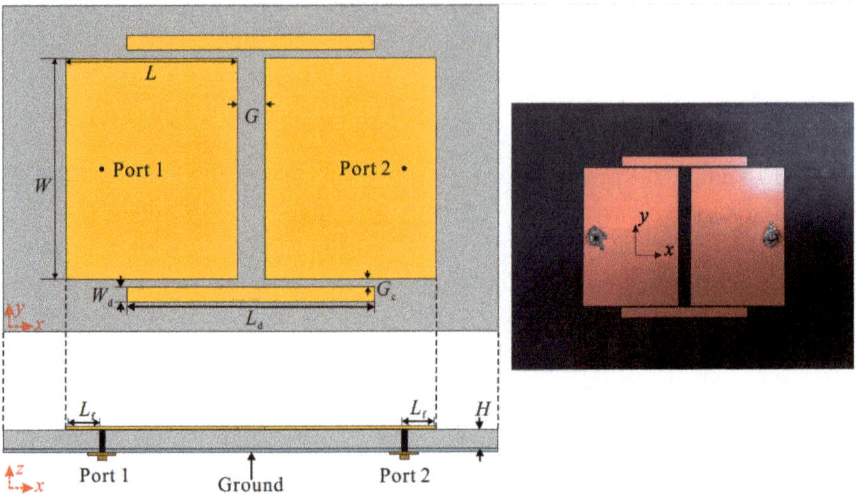

Figure 4.14 Schematic diagram and fabrication of the closely spaced MPAs with the proposed decoupling structure.

= 3.5 mm, G = 3 mm, L_d = 28.7 mm, W_d = 2 mm, G_c = 0.5 mm, L_f = 2.75 mm, and the relative permittivity of the substrate is $\varepsilon_r = 2.2$.

In order to depict the working principle about isolation enhancement, the corresponding analogous circuit model of the antenna in Figure 4.14 is plotted in Figure 4.15, which can be divided into the original coupling path (path A) and the additional decoupling path (path B). Meanwhile, the circuit model of two identical patch radiators are represented by two parallel *GLC* circuits, where the values of G_a, L_a, and C_a could be calculated under a single patch by using the extraction approach [17]. In addition, the admittance-inverter J_0 is adopted herein in order to represent the external coupling between the feeding ports and patch element, which is replaced by a parasitic series inductance L_p from the feeding probe. Moreover, for the mutual coupling between two radiated patches, it could be modeled as a capacitive π-network with the value of C_m. Obviously, C_m is the crucial parameter of this analogous circuit, which can be calculated by

$$C_m = CC\sqrt{C_a C_b} \tag{4.10}$$

where $C_a = C_b$, and *CC* is the coupling coefficient between the two radiating patches as can be calculated.

According to the dimensions of the patch radiators in Figure 4.14, the corresponding lumped elements are calculated to be G_a = 7.41 mS, L_a = 0.48 nH, C_a = 3.28 pF, and L_p = 2.59 nH. C_m is easily determined to be 0.28 pF. As such, the calculated S-parameters of path A can be derived and calculated.

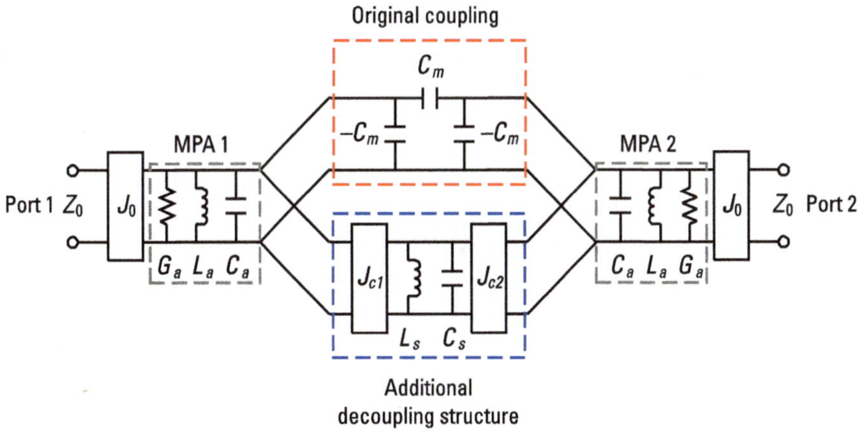

Figure 4.15 Corresponding analogous model at the circuit level of the entire structure in Figure 4.14.

Except for the path A, the analogous circuit of the additional coupling path (path B) could be also calculated. Herein, as observed in Figure 4.15, a parallel *LC* circuit is used to model the open-circuit half-wavelength microstrip line, whose values (L_s and C_s) can be calculated based on the closed-form expressions in [18]. In addition, the admittance-inverters J_{c1} and J_{c2} represent the coupling extent between two radiating patches and nonradiative half-wave microstrip line, respectively, which could be detail calculated based on the previous work [19, 20].

By using the above simulated results, the proposed antenna in Figure 4.14 could be designed, fabricated, and measured, and the corresponding calculated, simulated, and measured S-parameters of this designed antenna are given in Figure 4.16. It shows that the S_{21} under circuit model from Figure 4.16 agrees well with that of the full-wave simulated results. Besides, the measured results are basically consistent with the simulated results, thus demonstrating an impedance matching bandwidth ($S_{11} < -10$ dB) with the range of 4.38 to 4.60 GHz (4.9%). Compared with the initial isolation of above 7 dB, the maximum isolation is improved to 58 dB.

All in all, the coupling-interface technique based on open-ended microstrip stripline has been demonstrated as a useful method for isolation improvement of the small array.

4.3.3 Coupling Interface from Short-Circuited Microstrip Stripline

Apart from a single band with high-isolation performance, the coupling-interface technique can be extended to the dual-band isolation enhancement. As show in Figure 4.17, two proposed dual-frequency patch antennas (DFPAs) are

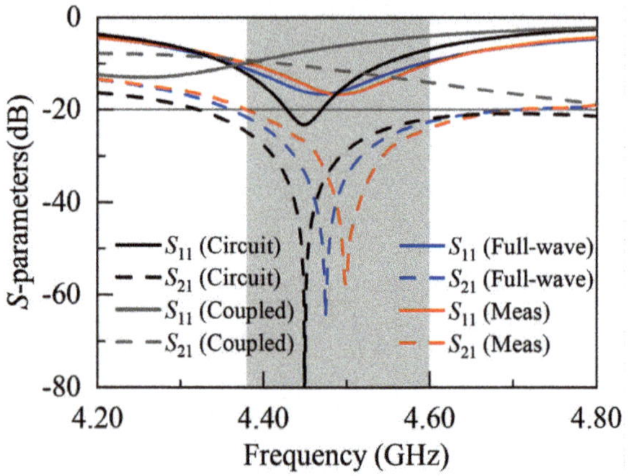

Figure 4.16 Calculated, simulated, and measured S-parameters of the proposed closely spaced MPAs and the coupled case.

placed symmetrically about the yoz-plane so that the naturally occurring pair of hybrid coupling interface (HCIs) can be directly connected [21]. The detail parameters of the antenna in [21] are listed as follows: $W = 30$ mm, $L = 21.5$ mm, $W_s = 3$ mm, $L_s = 22.3$ mm, $R = 0.8$ mm, $L_f = 6.5$ mm, $H_s = 0.8$ mm, $G = 3$ mm, and $H = 3.5$ mm, and the relative permittivity of the substrate is $\varepsilon_r = 2.2$. Herein, two shorting pins are simplified to share a common one. All dimensions are consistent with the single DFPA, except that the spacing between the two antennas is $G = 3$ mm ($0.039\lambda_L$). Besides, another group of two DFPAs is placed back-to-back at the same spacing without utilizing the HCI as a reference for the coupled case.

In order to illustrate its working principle, the equivalent circuit of the two-element array is modeled and plotted in Figure 4.18. Herein, J_E and J_M represent the coupling strength between its electric and magnetic coupling interface (CI), respectively. With respect to the value of J_E, it is determined by the spacing G, which could be extracted from two independent radiating patches under weak external coupling. With respect to the value of J_M, it is determined by the pins radius R, which could also be extracted. For this designed antenna, J_E should be kept about 6.28 mS since this value reflects the undesired mutual coupling between two radiating elements, meanwhile J_M can be manipulated to achieve self-decoupling in the desired dual-frequency operation.

The proposed antenna was designed, fabricated, and measured according to the structure shown in Figure 4.17. The simulated and measured S-parameters of the proposed antenna are shown in Figure 4.19. The measured results for two DFPAs with and without using the proposed HCI technique agree well with the simulated ones. Both in-band isolations are better than 15 dB, with

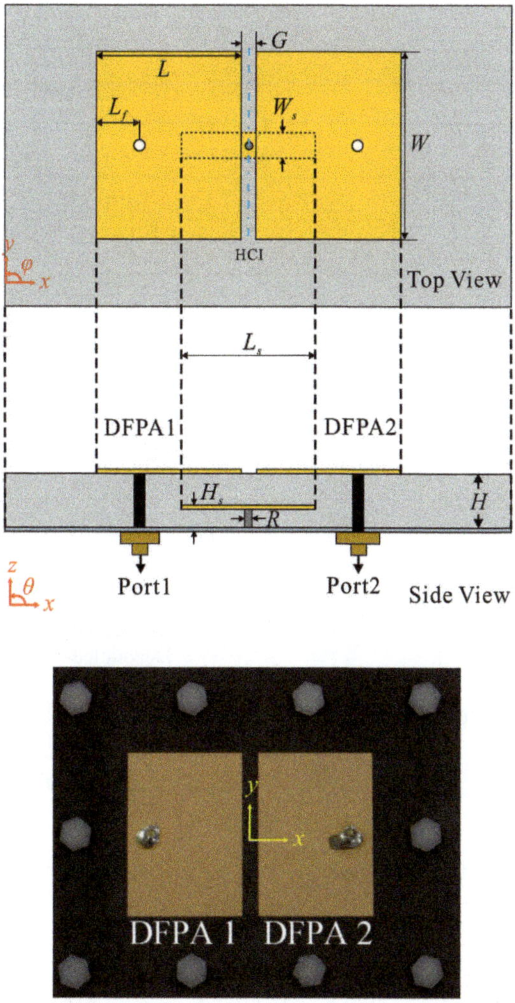

Figure 4.17 Configurations and fabrication of the proposed DFPAs with the decoupling shorted-circuited microstrip stripline.

good impedance matching bandwidths of $|S_{11}| < -10$ dB cover 3.77 to 3.94 GHz (4.4%), and 4.59 to 4.72 GHz (2.8%), respectively.

The proposed hybrid coupling-interface technique based on short-circuited microstrip stripline can reduce the mutual coupling between two DFPAs.

In addition to the three decoupling designs via coupling interfaces mentioned above, some other similar designs are listed, as shown in Figure 4.20. In [22], by introducing three interdigital lines with a coupled-mode transformation effect between patch antennas, orthogonally polarized modes can be generated on adjacent patches, resulting in an isolation better than 20 dB. In [23],

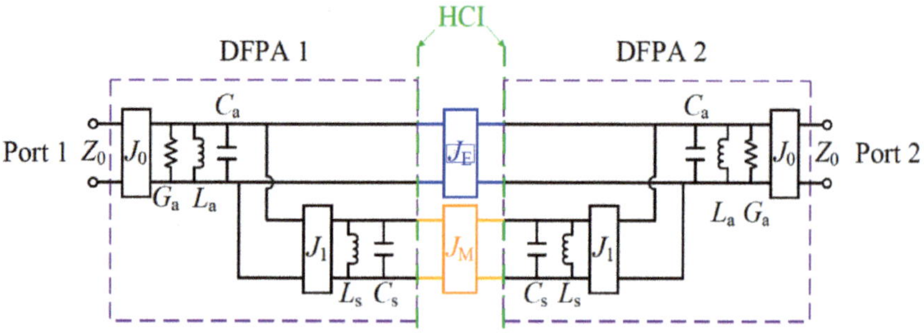

Figure 4.18 Corresponding equivalent-circuit model of the antennas in Figure 4.17.

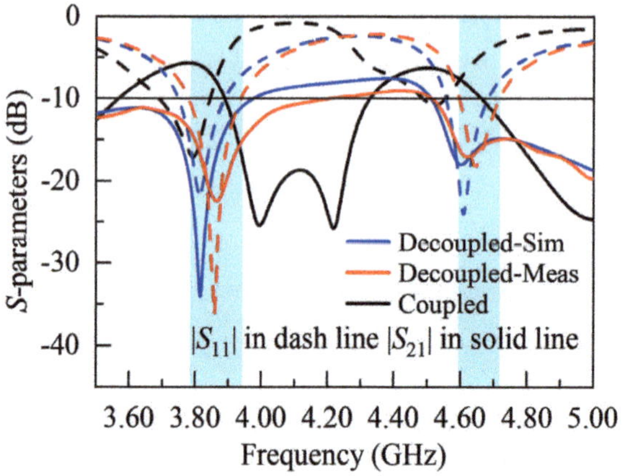

Figure 4.19 Simulated and measured *S*-parameter for the proposed antenna with and without the decoupling structure.

the decoupling unit consists of an asymmetric loop resonator with coupled microstrip lines to achieve a wide bandwidth from 2 to 5 GHz, providing greater than 15 dB of additional isolation in dual-band microstrip antenna arrays. This is in contrast to the above two methods, where coupling interfaces are added between the antennas. In [24], decoupling is achieved by adding coupling interfaces to the feed network. By introducing an open-circuit T-stub between the feed networks, the mutual coupling of a two-element E-plane antenna array is reduced by 24.2 dB and can be extended to a multivariate antenna array.

4.4 Mode-Combination Technique

The above techniques in Sections 4.2 and 4.3 improve isolation for two or more separated antennas. In this part, for a single multiport antenna, we discuss the working principle of mode-combination techniques for isolation enhancement.

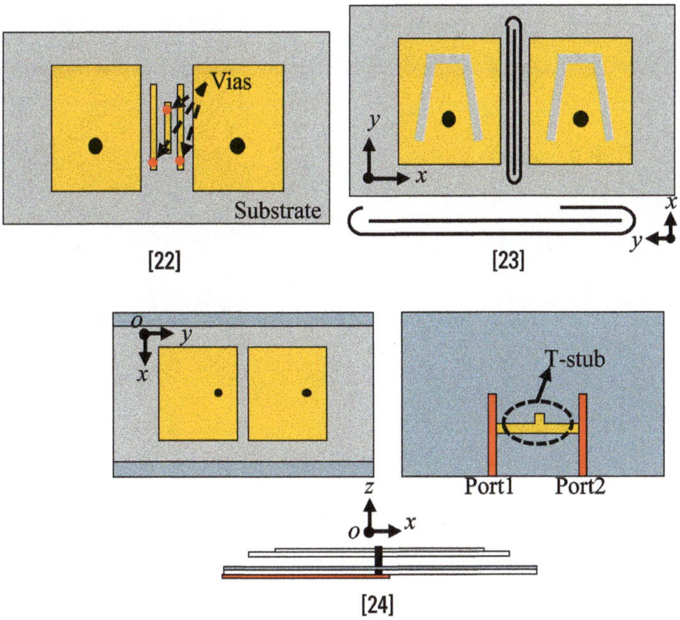

Figure 4.20 Other designs that decouple by coupling interface.

It mainly consists of mode combination in a single operating band in Section 4.4.1, mode combination in dual operating bands in Section 4.4.2, and mode combination in wide operating band in Section 4.4.3.

4.4.1 Mode Combination in a Single Operating Band

Figure 4.21 presents the configuration and fabrication of our proposed rectangular MPA. It consists of the radiating patch, six shorting pins, two feeding ports, and circular loop, ground plane, and dielectric substrate [25]. The detail parameters of the antenna are: $L = 24$, $L_g = 60$, $W = 25.4$, $W_g = 50$, $D_1 = 10.6$, $D_2 = 10.1$, $D_3 = 4.4$, $D_4 = 12.7$, $S_1 = 5.4$, $S_2 = 2.6$, $R_1 = 3.8$, $R_2 = 2.2$, $R_3 = 2.3$, $R_4 = 2.18$, $R_5 = 1.6$, $H = 4$. Unit: mm. We discuss the detailed analysis process and working principle in the following.

For the MPA in low-profile condition, the thickness (H) of the substrate is electrically smaller than the free-space wavelength ($H<<\lambda_0$). A set of modes could be excited in the wide operation band. In this context, only mode 2 (TM_{01}) and mode 4 (TM_{20}) are used and combined for isolation improvement. Initially, based on a classical cavity model [26, 27], the electric-field distributions of a low-profile MPA at mode 2 and mode 4 under the excitation of port 1 could be calculated as

$$E_{mode\,2} = B_l \cos(\pi y / L) \tag{4.11}$$

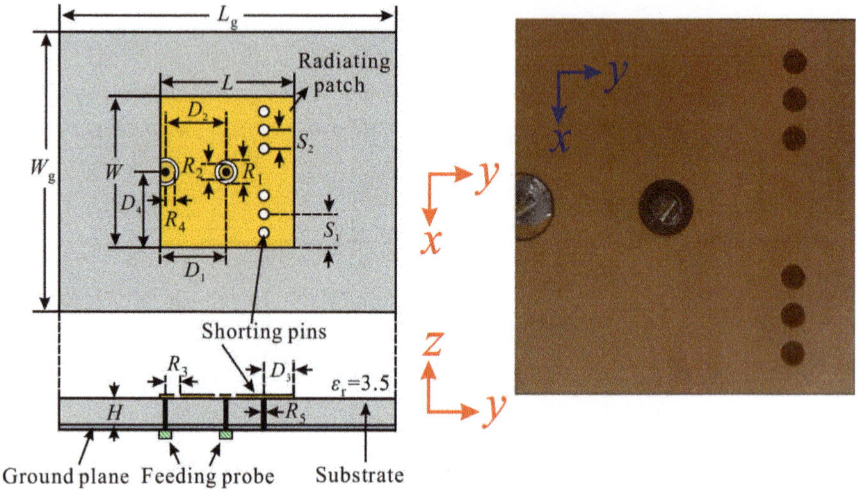

Figure 4.21 Configuration and fabrication of our proposed MPA with loaded pins.

$$E_{mode\ 4} = B_2 \cos(2\pi x / W) \tag{4.12}$$

where B_1 and B_2 refers to the amplitude of the electric field of mode 2 and mode 4, respectively.

After that, if we assume the magnitude ratio of the electric fields between mode 2 and mode 4 as n, the combined electric-field distributions under excitation of port 1 could be derived as

$$\begin{aligned} E_{total} &= E_{mode\ 2} + E_{mode\ 4} \\ &= B_1 \cos(\pi y / L) + B_2 \cos(2\pi x / W) \\ &= \begin{cases} B_1 \cos(\pi y / L) + B_1 / n \times \cos(2\pi x / W) & n \geq 1 \\ n \times B_2 \cos(\pi y / L) + B_2 \times \cos(2\pi x / W) & n < 1 \end{cases} \end{aligned} \tag{4.13}$$

By using (4.13), the calculated nodal lines of the MPA under two different ports are discussed in Figure 4.22. When port 1 is excited and port 2 is unexcited, the nodal lines variation under different n are plotted in Figure 4.22(a). Initially, as $n = 0$, only the electric field of mode 4 could be excited, which is based on (4.12). Two parallel nodes at $x = 0.25W$ and $x = 0.75W$ are formed up. By increasing n to above 1, the nodes are gradually bent in trajectory to generate an intersection point on the right-side patch edge, which is reallocated in proximity to the position of unexcited port 2. The intersection point refers to the point across the electric-field node of E_{total} and the central line of $x = 0.5W$. After that, the intersection point is gradually moved to the patch center

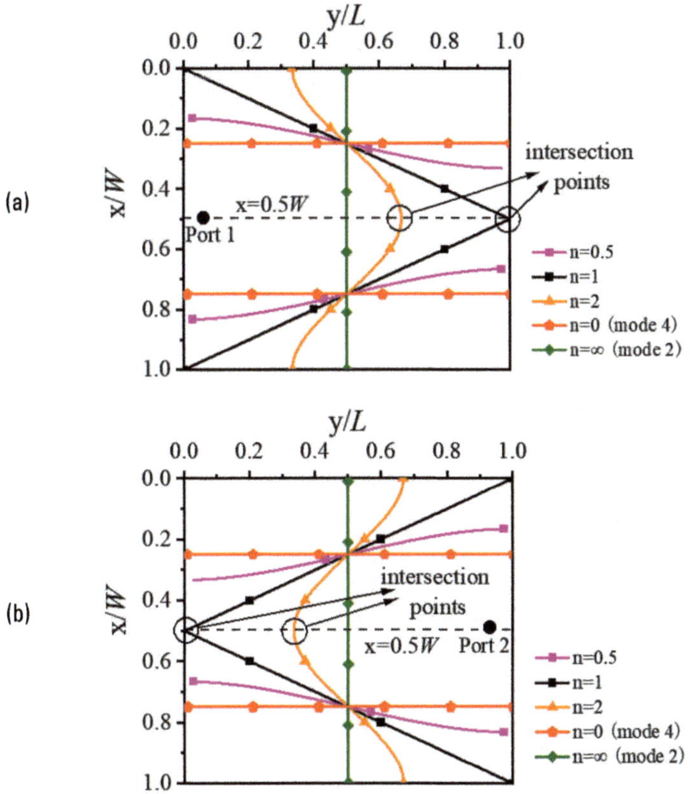

Figure 4.22 Nodal lines of the MPA by combining its mode 2 and mode 4 as a function of n under different ports: (a) port 1 excitation, and (b) port 2 excitation.

by further increasing n. As n approaches infinity, only mode 2 could be excited, leading to the production of a node line at $y = 0.5L$ from (4.11).

In contrast, when port 2 is excited and port 1 is unexcited, the electric-field distribution of mode 2 keeps the 180°-phase difference as compared to the results in (4.11). However, the electric-field distribution of mode 4 in this condition remains almost constant with that of (4.12). If we assume the magnitude ratio of the electric fields between mode 2 and mode 4 as n, the combined electric-field distributions from these dual-combined modes under excitation of port 2 could be expressed as

$$\begin{aligned} E_{total} &= E_{mode\,2} + E_{mode\,4} \\ &= B_1 \cos(\pi y / L + \pi) + B_2 \cos(2\pi x / W) \\ &= \begin{cases} B_1 \cos(\pi y / L + \pi) + B_1 / n \times \cos(2\pi x / W) & n \geq 1 \\ n \times B_2 \cos(\pi y / L + \pi) + B_2 \times \cos(2\pi x / W) & n < 1 \end{cases} \end{aligned} \tag{4.14}$$

By using (4.14), the calculated nodal lines of the antenna at port 2 under different n have been studied extensively in Figure 4.22(b). As $n = 0$, only the electric fields of the mode 4 could be excited. It generates parallel nodes at $x = 0.25W$ and $x = 0.75W$ based on (4.12). By increasing the magnitude ratio n, the nodal line is gradually bent to form the intersection point on the counterpart left-side edges, which is reallocated in proximity to the unexcited port 1. As n approaches infinity, only mode 2 can be excited to form up a single node line at $y = 0.5L$. The above calculated results show that the electric-field nodes of the antenna could be bent by controlling the magnitude ratio between these two combined modes, forming up the right- and left-intersection points, respectively. Meanwhile, if the unexcited port is located around the intersection points of these nodal lines, an antenna with high isolation could be generated between two ports.

To verify these predicted electric-field node-bending actions, the simulated results of the MPA under excitation of port 1 and port 2 independently are plotted in Figure 4.23(a, b), respectively. The feeding port 1 and port 2 are both located at $x = 0.5W$. Two types of electric-field node are satisfactorily obtained around the unexcited port 1 and port 2 independently, which is similar to the calculated results in Figure 4.22. The desired low mutual coupling under dual ports is effectively gained for the proposed antenna. Six shorting pins are loaded to combine mode 2 and mode 4, gaining the desired magnitude ratio n between these electric fields.

To validate the predicted performance, a proposed MPA with the loaded pins was fabricated and measured. The performance of the proposed MPA is compared here with that of a traditional antenna with two feeding ports under TM_{01} mode. Both $|S_{11}|$ and $|S_{22}|$ of the antenna are kept below –10 dB from 5.30 to 5.54 GHz. The MPA has low-profile ($0.07\lambda_0$), single-layer and

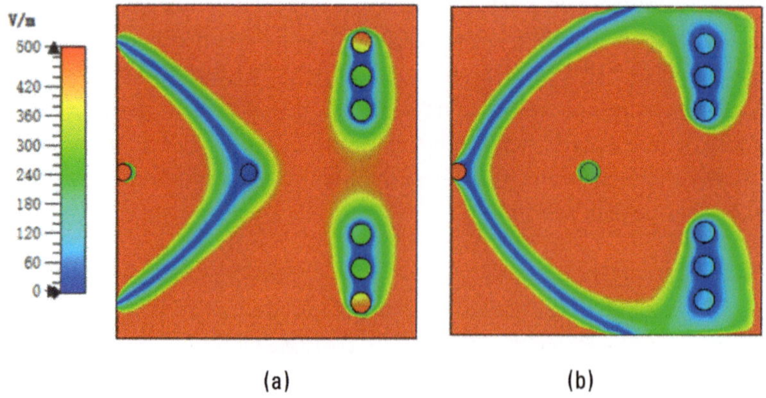

Figure 4.23 Electric-field distributions of the MPA loaded with six pins under different ports at 5.4 GHz: (a) port 1 excitation, and (b) port 2 excitation.

single-element properties as desired. The $|S_{12}|$ of the dual ports is kept below –17 dB, which is much smaller than the –3.5 dB of the traditional method under operation of fundamental TM_{01} mode. Particularly, the same broadside radiation patterns are achieved at dual-feeding ports with colinear polarization of around 5.4 GHz. The design concept of multimode combination for isolation enhancement is proposed for same polarized pattern, especially with the sharing single radiator.

4.4.2 Mode Combination in Dual Operating Bands

The mode-combination method could be well applied for a multiport antenna in dual-band or even more bands operation. In the following, we illustrate this working principle in detail.

Figure 4.25 shows the configuration and fabrication of our proposed MPA [28]. The rectangular patch ($L_1 \times W_1$) and ground plane ($L_g \times W_g$) are placed on the top and bottom surfaces of the dielectric substrate, while shorting pins 1, 2, 3, and 4 are loaded underneath the radiating patch. The main target of shorting pins 1 and 2 is to improve isolation between the left and right feeding ports. The main target of shorting pins 3 and 4 is to improve isolation between the upper and lower feeding ports. The detail parameters of the antenna are: L_g

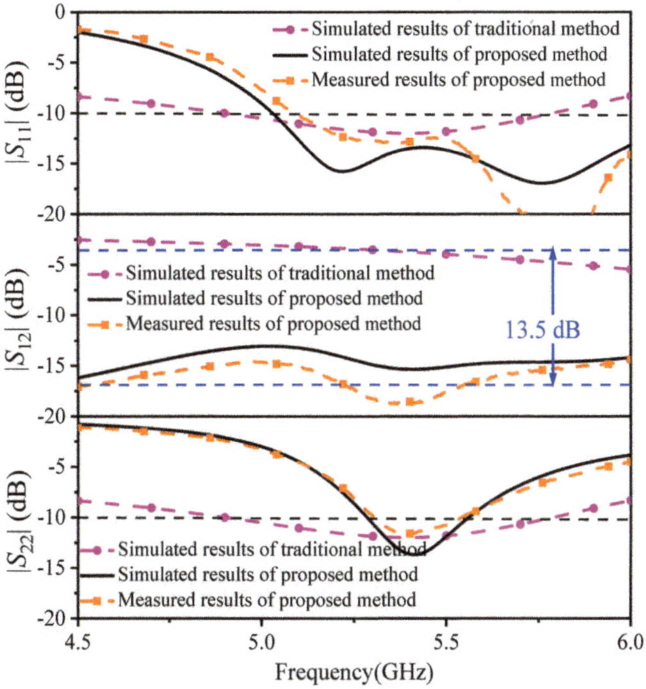

Figure 4.24 *S*-parameter comparison between the traditional and our proposed methods.

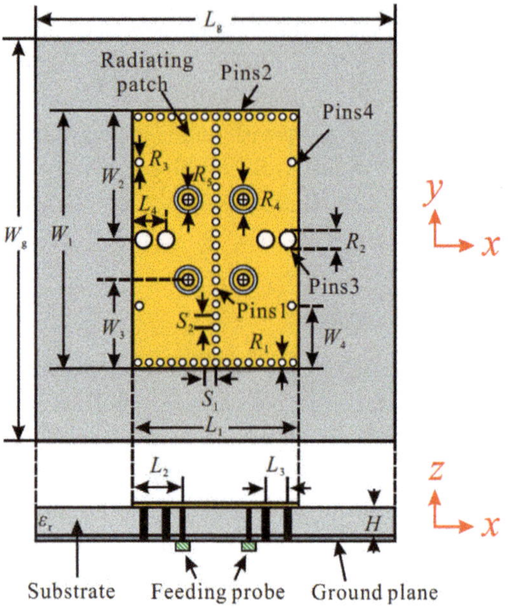

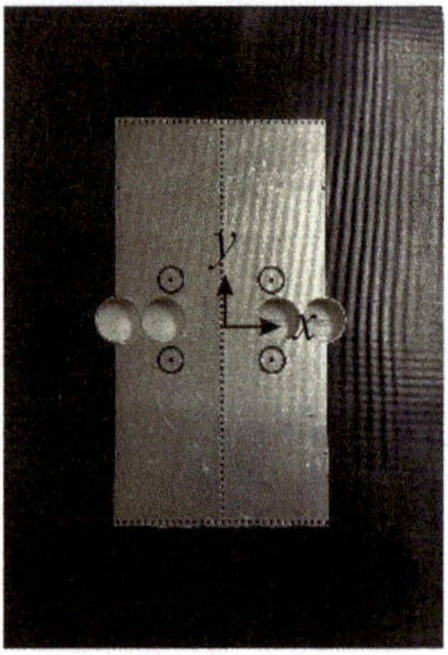

Figure 4.25 Configuration and fabrication of our proposed MPA with dual-band, high-isolation and copolarized pattern.

= 130, $L_1 = 60$, $L_2 = 15.7$, $L_3 = 13$, $L_4 = 13.7$, $W_g = 180$, $W_1 = 111$, $W_2 = 55.5$, $W_3 = 44.5$, $W_4 = 19$, $S_1 = 2$, $S_2 = 1.9$, $R_1 = 0.9$, $R_2 = 12$, $R_3 = 0.9$, $R_4 = 7.8$, $R_5 = 6.1$, $H = 7$. Unit: mm.

The isolation improvement between left and right feeding ports is due to shorting pins blocking between these internal electric fields. Our main target is improving isolation of the antenna between its lower and upper feeding ports, and we combine mode 1, mode 2, mode 3, and mode 4 to control nodal line distribution around the lower and upper bands. We investigate this working principle from two parts independently.

Part I: For the antenna at lower band, the electric-field distributions of mode 1 and mode 2 are theoretically studied by using the cavity model, and the corresponding schemes are illustrated in Figure 4.26. With respect to mode 1, its electric fields remain in the same direction underneath the patch, but the electric fields of mode 2 varied once along the *y*-axis, forming an electric node at the center of the radiating patch ($y = 0.5W_1$). The electric-field distribution of mode 1 and mode 2 under excitation of port 1 could be expressed as

$$E_{model\,1} = B_1 \sin(\pi \times y / W_1) \times \cos(\pi \times x / L_1) \tag{4.15}$$

$$E_{model\,2} = -B_2 \sin(2\pi \times y / W_1) \times \cos(\pi \times x / L_1) \tag{4.16}$$

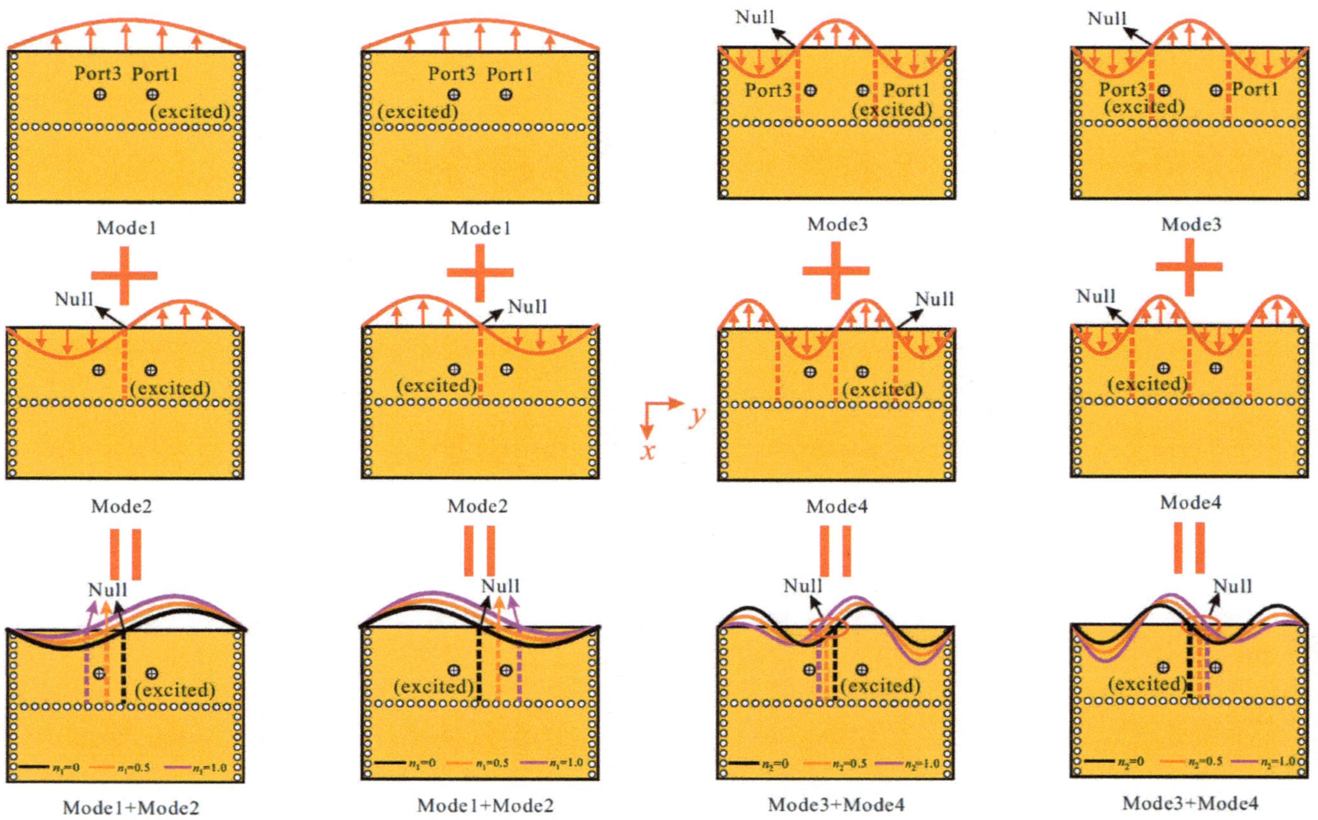

Figure 4.26 Electric field of the shorted MPA under different excited ports and different radiative modes.

where B_1 and B_2 refers to the amplitude of the electric field of mode 1 and mode 2, respectively.

After that, if we assume the magnitude ratio of the electric fields between mode 1 and mode 2 as n_1, the combined electric-field distributions from these dual-combined modes under excitation of port 1 should be calculated as

$$\begin{aligned} E_{total} = E_{model\ 1} + E_{model\ 2} = B_2(n_1 \times \sin(\pi \times y / W_1) \times \cos(\pi \times x / L_1) \\ - \sin(2\pi \times y / W_1) \times \cos(\pi \times x / L_1)) \end{aligned} \tag{4.17}$$

By using (4.17), the electric-field variation of the dual modes under excitation of port 1 and nonexcitation of port 3 is also investigated in Figure 4.26. The nodal line is kept at approximately $y = 0.5\,W_1$ at $n_1 = 0$, which means that only mode 2 is excited in this condition. Next, by increasing n_1 from 0 to 1.0, the nodal line from these combined modes is freely shifted to the left-side edge of the patch. The feeding port could be formed up around the unexcited feeding port 3.

In contrast, when port 3 is excited and port 1 is unexcited, the electric-field distribution of mode 1 is kept similar to that of the excited port 1, but the electric-field distribution of mode 2 is inversely directed compared to that of excited port 1. By increasing n_1 from 0 to 1.0, the combined nodal line from these two modes is shifted to the right-side edge of the radiating patch.

Based on the above theoretical investigation, the nodal line distribution of the MPA could be effectively controlled through the magnitude ratio between mode 1 and mode 2. When feeding port 1 and port 3 are placed around the corresponding nodal lines, isolation between these ports could be effectively increased. The symmetrical scheme means the mutual coupling toward port 2 and port 4 is also reduced in the lower area, which ignores the working principle discussion about isolation improvement. The magnitude ratio n_1 is mainly controlled by the frequency ratio between these modes, which could be realized by loading shorting pins 3 and 4.

Part II: For the antenna at upper band, the electric-field distributions of mode 3 and mode 4 are theoretically studied by using the cavity model, and the corresponding schemes are also illustrated in Figure 4.26. The electric fields of the antenna at mode 3 and mode 4 under excitation of port 1 could be derived as

$$E_{model\ 3} = -B_3 \sin(3\pi \times y / W_1) \times \cos(\pi \times x / L_1) \tag{4.18}$$

$$E_{model\ 4} = B_4 \sin(4\pi \times y / W_1) \times \cos(\pi \times x / L_1) \tag{4.19}$$

where B_3 and B_4 refers to the amplitude of the electric field of mode 3 and mode 4, respectively.

After that, if we assume the magnitude ratio of the electric fields between mode 3 and mode 4 as n_2, the combined electric-field distributions from these dual-combined modes under excitation of port 1 should be calculated as

$$\begin{aligned} E_{total} &= E_{model\ 3} + E_{model\ 4} \\ &= -B_4(n_2 \times \sin(3\pi \times y / W_1) \times \cos(\pi \times x / L_1) \\ &- \sin(4\pi \times y / W_1) \times \cos(\pi \times x / L_1)) \end{aligned} \tag{4.20}$$

By using (4.20), the combined electric field from mode 3 and mode 4 under excitation of port 1 and nonexcitation of port 3 is also presented in Figure 4.26. Similar to the previous tendency, the nodal line is freely moved to the left-side edge by increasing n_2 from 0 to 1.

In contrast, when port 3 is excited and port 1 is unexcited, the nodal line could be freely shifted to the right-side edge due to the 180°-phase difference of its mode 4 from (4.18). Using this method, the isolation of the antenna between port 2 and port 4 could also be enhanced due to the symmetrical scheme in the lower area. With these arrangements, the isolation of the antenna between its lower and upper ports is successfully enhanced via this method. Noticing that, the magnitude ratio n_2 is mainly controlled by the frequency ratio between these modes, which could be realized by loading shorting pins 3 and 4.

To further verify our predicted isolation improvement, the proposed antenna in Figure 4.25 with ring slots was manufactured and measured. Figure 4.27 plots the *S*-parameter of our four-port MPA. The MPA resonates at approximately 2.40 and 3.62 GHz in dual-operating bands. The simulated and measured mutual couplings ($|S_{12}|$, $|S_{13}|$, $|S_{14}|$, $|S_{23}|$, $|S_{24}|$, and $|S_{34}|$) of the MPA are also plotted. The mutual coupling between four ports is successfully kept as small as below –15 dB in these bands. The antenna still holds several advantages, such as low-profile, single-layer, and single-radiator properties without needing extra feeding networks. The copolarized normal patterns are satisfactorily produced in the dual bands at four ports, and the tilted beam and cross polarization of the MPA, especially in a single radiator, are caused by the asymmetric scheme. The design concept of multimode combination for isolation enhancement is proposed for the same polarized pattern, especially in dual working bands.

4.4.3 Mode Combination in Wide Operating Band

In this part, the mode-combination method is extended to the wide operating band for a multiport antenna. Figure 4.28 plots the structure and fabrication

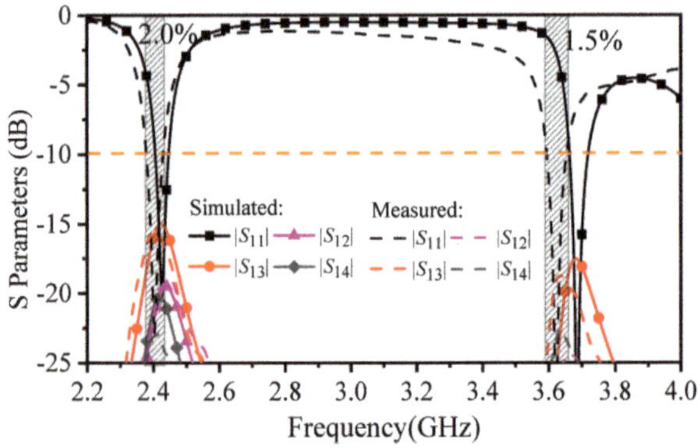

Figure 4.27 Simulated and measured S-parameter of the proposed MPA in Figure 4.25 loaded with ring slots.

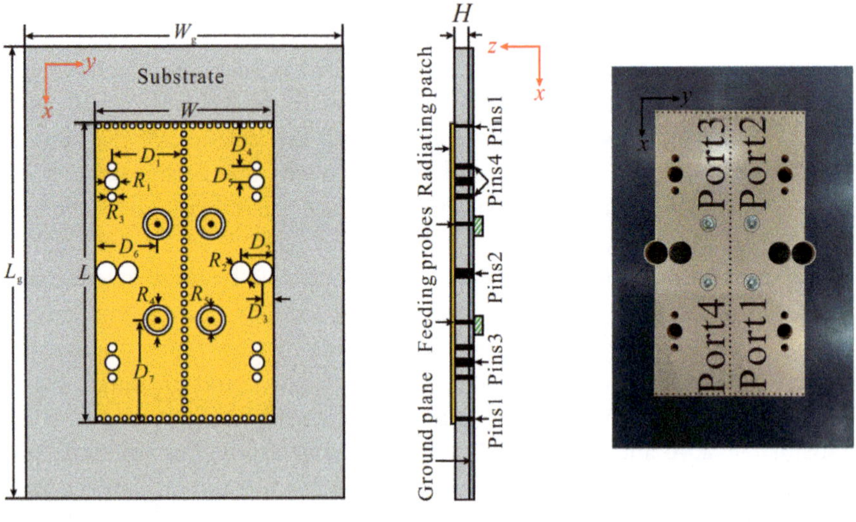

Figure 4.28 Configuration and fabrication of our proposed four-port antenna with wide high-isolation bandwidth.

of the proposed antenna [29]. The antenna consists of the radiating patch, dielectric substrate, shorting pins, ground plane, and slots. The detail parameters of the antenna are: $L_g = 150$, $L = 110$, $W_g = 100$, $2W = 60.6$, $R_1 = 6$, $D_1 = 22.4$, $D_2 = 10$, $R_2 = 9$, $D_3 = 0.5$, $D_4 = 18.5$, $R_3 = 3$, $D_5 = 6.2$, $D_6 = 21.35$, $D_7 = 43.5$, $R_4 = 6.8$, $R_5 = 6.3$, $H = 5$. Unit: mm.

Here, we use six modes of the traditional SPA, CM1, CM2, CM3, CM4, CM6, and CM7. To realize wideband isolation bandwidth, the combined two modes for isolation improvement should satisfy two conditions: (1) the electric fields of these combined modes should be kept consistent along one direction with commonly used nodes; and (2) the electric fields of these combined modes should be varied odd times and even times respectively along another direction. According to this principle, we should be able to combine the electric fields of CM1, CM3, and CM6 with those of CM2, CM4, and CM7. The combination process of different modes will be further studied in the following three parts.

Part I: First, we discuss electric-field variation under CM1 and CM2. Based on previous works [30, 31], for the antenna under excitation of port 1 and nonexcitation of port 2, the internal electric fields from CM1 and CM2 could be expressed as

$$E_{CM1} = A_1 \sin(\pi \times y / 2W) \tag{4.21}$$

$$E_{CM2} = A_2 \cos(\pi \times x / L) \times \sin(\pi \times y / 2W) \tag{4.22}$$

where A_1 and A_2 are the amplitudes of these internal electric fields at CM1 and CM2, respectively.

After that, if we assume the magnitude ratio of the electric fields between CM1 and CM2 as n_1, the combined electric-field distributions from these dual-combined modes under excitation of port 1 should be derived as

$$\begin{aligned} E_{CM1+CM2} &= n_1 \times A_2 \sin(\pi \times y / 2W) \\ &+ A_2 \cos(\pi \times x / L) \times \sin(\pi \times y / 2W) \end{aligned} \tag{4.23}$$

Using (4.21) to (4.23), the electric-field distributions before and after combining CM1 and CM2 are discussed in Figure 4.29. By combining the two modes and setting n_1 to 0.6, the node is moved to $x = 0.70L$ of the right-side edge. Due to the symmetrical structure, the node line could be moved to the left-side patch edge under excitation of port 2 and nonexcitation of port 1. If the excited port is located far away from the node with the unexcited port near this place, the isolation between two ports could be enhanced with a pole in the frequency response.

Part II: Similar to the above method, for the antenna under excitation of port 1 and nonexcitation of port 2, the electric field of the SPA at CM3 and CM4 could be set as

$$E_{CM3} = A_3 \cos(2\pi \times x / L + \pi) \times \sin(\pi \times y / 2W) \tag{4.24}$$

$$E_{CM4} = A_4 \cos(3\pi \times x / L + \pi) \times \sin(\pi \times y / 2W) \tag{4.25}$$

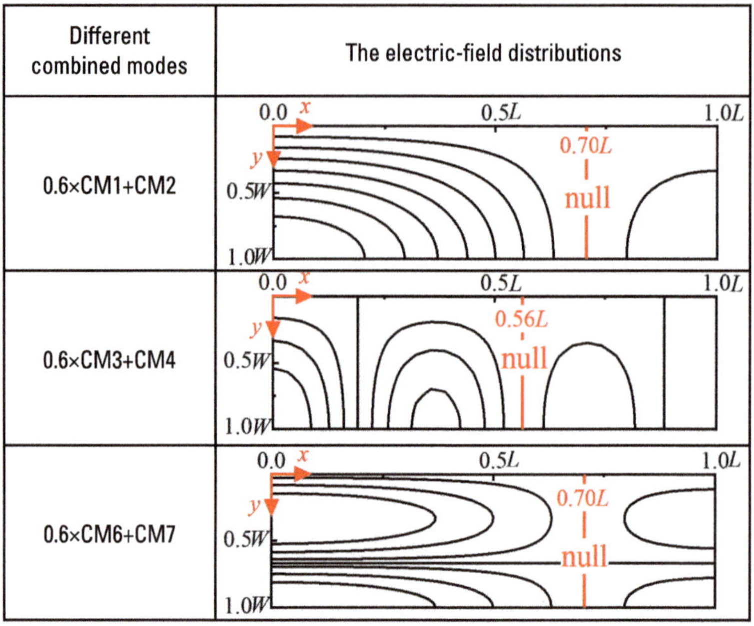

Figure 4.29 Contour of the electric-field distributions under different combinations of CM1, CM2, CM3, CM4, CM6, and CM7.

where A_3 and A_4 are the amplitudes of these internal electric fields at CM3 and CM4, respectively.

After that, if we assume the magnitude ratio of the electric fields between CM3 and CM4 as n_2, the combined electric-field distributions from these dual-combined modes under excitation of port 1 should be derived as

$$\begin{aligned} E_{CM3+CM4} &= n_2 \times A_4 \cos(2\pi \times x / L + \pi) \times \sin(\pi \times y / 2W) \\ &+ A_4 \cos(3\pi \times x / L + \pi) \times \sin(\pi \times y / 2W) \end{aligned} \tag{4.26}$$

From (4.24) to (4.26), the electric-field distributions of different conditions are also plotted and discussed in Figure 4.29. It shows that the nodal line is located at $x = 0.56L$ by setting n_2 as 0.6 so that the second minimum pole of its MC could be successfully gained. Due to the symmetrical structure, the node line could be moved to the left-side patch edge under excitation of port 2 and nonexcitation of port 1. If the excited port is located far away from the node with the unexcited port near this place, the second isolation pole could be generated in the wideband as desired.

Part III: With CM6 and CM7, for the antenna under excitation of port 1 and nonexcitation of port 2, the electric-field expressions of these two modes are obtained as

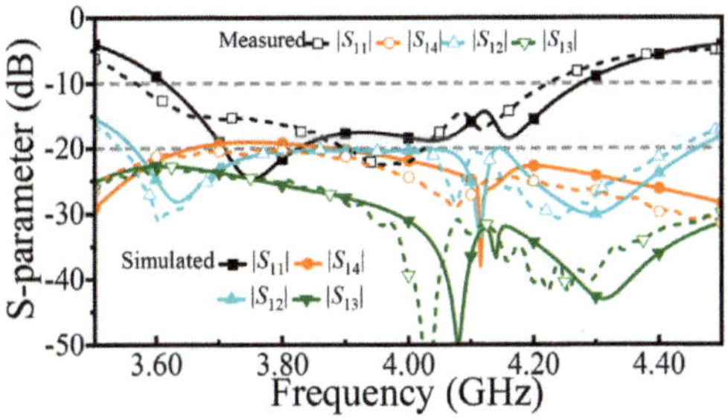

Figure 4.30 Simulated and measured S-parameter of our proposed MPA with wideband high-isolation bandwidth in Figure 4.28.

$$E_{CM6} = A_6 \sin(3\pi \times y / 2W) \tag{4.27}$$

$$E_{CM7} = A_7 \cos(\pi \times x / L) \times \sin(3\pi \times y / 2W) \tag{4.28}$$

where A_6 and A_7 are the amplitudes of these internal electric fields at CM6 and CM7, respectively.

After that, if we assume the magnitude ratio of the electric fields between CM6 and CM7 as n_3, the combined electric-field distributions from these dual-combined modes under excitation of port 1 should be derived as

$$\begin{aligned} E_{CM6+CM7} &= n_3 \times A_7 \sin(3\pi \times y / 2W) \\ &+ A_7 \cos(\pi \times x / L) \times \sin(3\pi \times y / 2W) \end{aligned} \tag{4.29}$$

If we set the value of n_3 as 0.6, the nodal line of the combined CM6 and CM7 is located at $x = 0.70L$. Due to the symmetrical structure, the node line could be moved to the left-side patch edge under excitation of port 2 and non-excitation of port 1. In this case, the third pole for the isolation could be gained via this method in the wideband.

The magnitude ratios n_1, n_2, and n_3 are mainly controlled by the frequency ratio between these modes, which could be realized by loading the shorting pins 3 and 4.

Figure 4.30 shows the S-parameter of our proposed four-port MPA. The measured results indicate that the isolation bandwidth of above 20 dB is improved to about 16.2% from 3.575 to 4.205 GHz. The antenna keeps the stable broadside radiation patterns. The design concept of multimode combination for a single MPA to improve isolation between four ports is proposed, especially in the wide working band with same polarized patterns.

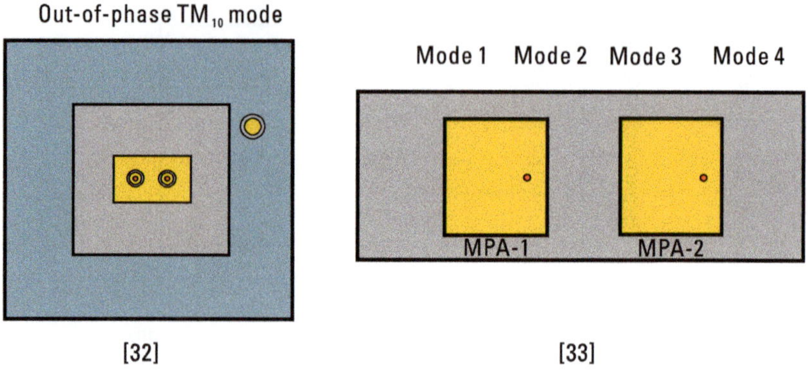

Figure 4.31 Other designs that decouple by mode combination.

In addition, some designs also use the mode-combination method, as shown in Figure 4.31. In [32], two out-of-phase TM_{10} modes are excited to achieve the low MC in the wideband, especially at high thickness. In [33], by adjusting the frequencies of two modes, the *E*-fields coupled to adjacent passive MPA can counteract each other, creating a specific zero-field region at the feed, resulting in an isolation of better than 51 dB. In [34, 35], several modes of the antennas could be combined, improving the isolation to more than 30 dB.

4.5 Conclusions

The isolation bandwidth of antennas may be enhanced by using DGSs, neutralization lines, split-ring resonators, slotted meander-line resonators, and so on. But compared to these abovementioned methods, the multimode resonance-inspired antennas with high isolation in this section have the advantages of compact size, wide bandwidth, or copolarized pattern. The proposed antennas have potential applications in several MIMO communication systems, such as smartphone devices, in-band full-duplex systems, phased arrays, and so on.

References

[1] M. -A. Jensen, and J. -W. Wallace, "A Review of Antennas and Propagation for MIMO Wireless Communications," *IEEE Transactions on Antennas and Propagation,* Vol. 52, No. 11, November 2004, pp. 2810–2824.

[2] Wong, K., C. Tsai, and J. Lu, "Two Asymmetrically Mirrored Gap-Coupled Loop Antennas as a Compact Building Block for Eight-antenna MIMO Array in the Future Smartphone," *IEEE Transactions on Antennas and Propagation,* Vol. 65, No. 4, April 2017, pp. 1765–1778.

[3] Lin, J., and Q. Chu, "Accurately Characterizing the Coupling Effects of Patch Antennas with Complex- and Frequency-Dependent J/K Inverters," *IEEE Transactions on Antennas and Propagation,* Vol. 67, No. 3, March 2019, pp. 1554–1561.

[4] A. Derneryd, "A Theoretical Investigation of the Rectangular Microstrip Antenna Element," *IEEE Transactions on Antennas and Propagation,* Vol. 26, No. 4, July 1978, pp. 532–535.

[5] Pozar, D. -M., *Microwave Engineering,* Hoboken, NJ: Wiley, 2011, pp. 421–422.

[6] Hong, J. -S., and M. -J. Lancaster, "Couplings of Microstrip Square Open-Loop Resonators for Cross-Coupled Planar Microwave Filters," *IEEE Transactions on Microwave Theory and Techniques,* Vol. 44, No. 11, November 1996, pp. 2099–2109.

[7] Hong, J.-S., and M. -J. Lancaster, *Microstrip Filters for RF/Microwave Applications,* Hoboken, NJ: Wiley, 2011, pp. 202–209.

[8] Zeng, Q. -Y., et al., "Decoupling of Antenna Pairs Based on Equal Modal Conductance by Antenna-Shape Modification," *IEEE Transactions on Antennas and Propagation,* Vol. 71, No. 3, March 2023, pp. 2182–2193.

[9] Cheng, Y. -F., and K. -K. M. Cheng, "Decoupling of Two-element Printed-Dipole Antenna Array by Optimal Meandering Design," *IEEE Transactions on Antennas and Propagation,* Vol. 68, No. 11, November 2020, pp. 7328–7338.

[10] Chen, Y., and C.-F. Wang, *Characteristic Modes: Theory and Applications in Antenna Engineering,* Hoboken, NJ: Wiley, 2015, pp. 1–95.

[11] Cornelius, R., et al., "Calculating the Envelope Correlation Coefficient Directly from Spherical Modes Spectrum," *11th European Conference on Antennas and Propagation (EUCAP),* Paris, France, May 2017, pp. 3003–3006.

[12] Sun, L., et al., "Antenna Decoupling by Common and Differential Modes Cancellation," *IEEE Transactions on Antennas and Propagation,* Vol. 69, No. 2, February 2021, pp. 672–682.

[13] Sun, L., et al., "Self-Decoupled MIMO Antenna Pair with Shared Radiator for 5G Smartphones," *IEEE Trans. Antennas Propag.*, Vol. 68, No. 5, May 2020, pp. 3423–3432.

[14] Sun, L., Y. Li, and Z. Zhang, "Decoupling Between Extremely Closely Spaced Patch Antennas by Mode Cancellation Method," *IEEE Transactions on Antennas and Propagation,* Vol. 69, No. 6, June 2021, pp. 3074–3083.

[15] Pei, T., et al., "A Low-Profile Decoupling Structure for Mutual Coupling Suppression in MIMO Patch Antenna," *IEEE Transactions on Antennas and Propagation,* Vol. 69, No. 10, October 2021, pp. 6145–6153.

[16] Deng, H., L. Zhu, and Z. -A. Ouyang, "Systematic Design Method for Mutual Coupling Reduction in Closely Spaced Patch Antennas," *IEEE Open Journal of Antennas and Propagation,* Vol. 4, 2023, pp. 349–360.

[17] Jiang, Z. -H., and D. -H. Werner. "A Compact, Wideband Circularly Polarized Codesigned Filtering Antenna and its Application for Wearable Devices with Low SAR," *IEEE Transactions on Antennas and Propagation,* Vol. 63, No. 9, September 2015, pp. 3808–3818.

[18] Gupta, K. -C., R. Garg, and I. -J. Bahl, *Microstrip Lines and Slotlines,* Norwood, MA: Artech House, 1979.

[19] Wu, Q.-S., X. Zhang, and L. Zhu, "An Improved Method for Accurate Extraction of Coupling Coefficient Between a Lossy Radiator and a Lossless Resonator in Filtering Antennas," *IEEE Access,* Vol. 6, 2018, pp. 39927–39935.

[20] Zhang, S., and L. Zhu, "Synthesis Method for Even-Order Symmetrical Chebyshev Bandpass Filters with Alternative J/K Inverters and Lambda/4 Resonators," *IEEE Transactions on Microwave Theory and Techniques,* Vol. 61, No. 2, February 2013, pp. 808–816.

[21] Deng, H., and L. Zhu, "Self-Decoupled Dual-Frequency Patch Antennas Via Hybrid Coupling Interface Technique," *IEEE Antennas and Wireless Propagation Letters,* Vol. 22, No. 6, June 2023, pp. 1256–1260.

[22] Qi, H., et al., "Improving Isolation Between Closely Spaced Patch Antennas Using Interdigital Lines," *IEEE Antennas and Wireless Propagation Letters,* Vol. 15, 2016, pp. 286–289.

[23] Dhevi, B. L., et al., "Isolation Enhancement in Dual-Band Microstrip Antenna Array Using Asymmetric Loop Resonator," *IEEE Antennas and Wireless Propagation Letters,* Vol. 17, No. 2, February 2018, pp. 238–241.

[24] Zou, X. J., et al., "An Efficient Decoupling Network Between Feeding Points for Multielement Linear Arrays," *IEEE Transactions on Antennas and Propagation,* Vol. 67, No. 5, May 2019, pp. 3101–3108.

[25] Liu, N. -W., et al., "Electric-Field Null Bending of a Single Dual-Port Patch Antenna for Colinear Polarization Decoupling Using Characteristic Modes Analysis," *IEEE Transactions on Antennas and Propagation,* Vol. 70, No. 12, December 2022, pp. 12247–12252.

[26] Lo, Yo., D. Solomon, and W. Richards, "Theory and Experiment on Microstrip Antennas," *IEEE Transactions on Antennas and Propagation,* Vol. 27, No. 2, March 1979, pp. 137–145.

[27] Boualleg, A., and N. Merabtine, "Analysis of Radiation Patterns of Rectangular Microstrip Antennas with Uniform Substrate," *Semiconductor Physics Quantum Electronics & Optoelectronics,* Vol. 8, No. 3, 2005, pp. 88–91.

[28] Liu, N. -W., et al., "Mutual Coupling Reduction of a Dual-Band Four-Port Patch Antenna with Co-polarized Radiation Pattern by Controlling Electric Fields," *IEEE Transactions on Antennas and Propagation,* Vol. 71, No. 5, May 2023, pp. 4534–4539.

[29] Liu, N. -W., et al., "Low-Profile Four-Port Patch Antenna with Wide Isolation Bandwidth and Same Polarization: Principle and Design Approach," *IEEE Antennas and Wireless Propagation Letters,* Vol. 22, No. 10, October 2023, pp. 2407–2411.

[30] Li, D., et al., "Dual-Band, Differentially-Fed Filtenna with Wide Bandwidth, High Selectivity, and Low Cross-Polarization," *IEEE Transactions on Antennas and Propagation,* Vol. 70, No. 6, June 2022, pp. 4872–4877.

[31] James, J. R., and P. S. Hall, *Handbook of Microstrip Antennas,* London, UK: Peregrinus, 1989.

[32] Wong, K. -L., et al, "Two-Port Same-Polarized Patch Antenna Based on Two Out-of-Phase TM_{10} Modes for Access-Point MIMO Antenna Application," *IEEE Antennas and Wireless Propagation Letters,* Vol. 20, No. 4, April 2021, pp. 572–576.

[33] Lai, Q. -X., et al., "Mutual Coupling Reduction in MIMO Microstrip Patch Array Using TM_{10} and TM_{02} Modes," *IEEE Transactions on Antennas and Propagation,* Vol. 69, No. 11, November 2021, pp. 7562–7571.

[34] Guo, C., et al., "Compact Co-Polarized Decoupled Microstrip Patch Array Antenna Based on TM_{02}/TM_{03} Modes Cancellation," *IEEE Transactions on Antennas and Propagation,* Vol. 70, No. 10, October 2022, pp. 9906–9911.

[35] Zha, L., Y. -M. Pan, and S. -Y. Zheng, "Self-Decoupled Linear and Planar MIMO Microstrip Patch Antenna Arrays Operating in the Fundamental TM_{01} Mode," *IEEE Transactions on Antennas and Propagation,* Vol. 72, No. 2, February 2024, pp. 1224–1233.

5

Multimode Resonance-Inspired Planar Antennas with Gain Bandwidth Improvement

5.1 Introduction

The main focus of this chapter is to show methods for improving the performance of planar antennas through gain bandwidth. Gain-enhancement techniques will be introduced in Section 5.2; stable-gain techniques will be discussed in Section 5.3; and CP bandwidth will be illustrated in Section 5.4.

5.2 High-Gain Bandwidth Technique

In modern wireless communication systems, high-gain antennas are effective for path-loss compensation and coverage enhancement. Unfortunately, traditional planar antennas suffer from relatively low gain. For example, the gain of a printed MPA is only maintained at around 8 dBi or even lower. So far, several effective methods have been proposed and developed for these antennas during the past two decades, such as using shorting pins for fundamental mode in Section 5.2.1, using slots and shorting pins for fundamental mode in Section 5.2.2, and using slots and shorting pins for high-order modes in Section 5.2.3.

5.2.1 Shorting Pins for Fundamental Mode

The working principle for gain enhancement of a rectangular MPA operating at the fundamental TM_{01} mode will be introduced by using shorting pins [1]. The diagram in Figure 5.1 illustrates the proposed geometric layout of the patch antenna. It consists of a rectangular patch on an electrically thin substrate with a height (h) and dielectric constant (ε_r). Four metallic pins are positioned at the diagonals of the patch resonator to electrically short the patch and ground plane at the four corners of a rectangular contour. The patch is fed by a coaxial probe and operates under its dominant TM_{01} mode. The specific parameters of this designed antenna are: $W = 63$ mm, $D/W = 0.6$, $R = 4$ mm, the ground size is 170 mm × 170 mm, the relative permittivity (ε_r) of the dielectric substrate is 2.2, and its thickness (h) is 3.175 mm. The feed position d in each case is appropriately chosen to match a 50Ω probe [2]. The introduced loading of shorting pins will significantly perturb the fundamental mode of the antenna with increased electrical size, which will enhance the directivity of the antenna, as shown in the radiation pattern comparison in Figure 5.1.

The equivalent radiation model and far-field analysis of the proposed patch antenna have been discussed in detail in our reported work [1]. For brevity, only the high-gain mechanism is discussed here.

Suppose E_1 and E_2 represent the electric fields across the radiating slots. Next, according to Huygens' principle [2], the radiation patterns of the MPA operating in TM_{01} mode can be calculated by the classical double-slot array model. Assuming the electric field across each slot is uniformly distributed and the ground plane is infinite, the equivalent magnetic current, M_{s1} and M_{s2}, are given by

$$\overrightarrow{M_s} = \vec{E} \times \vec{n} \tag{5.1}$$

Resorting to aperture antenna theory [3], the far-field radiated by two magnetic currents at E-plane can be derived. The directivity of the antenna is finally expressed as

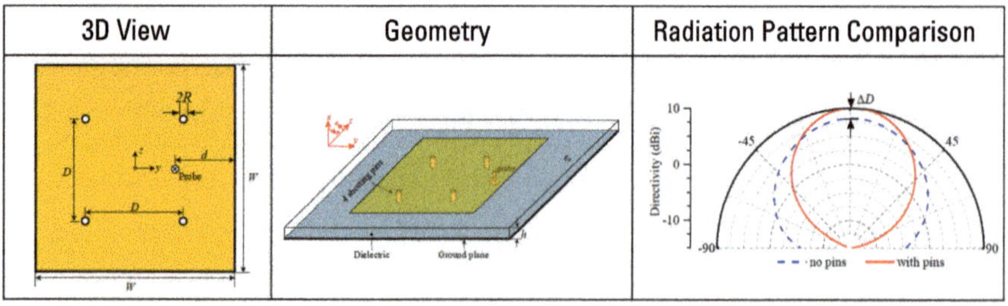

Figure 5.1 Geometry and radiation pattern of the proposed patch antenna with four shorting pins.

$$D_t = \frac{4\pi U_{\max}}{P_{rad}} = (\frac{2\pi W}{\lambda})^2 \frac{\pi}{I_2} \tag{5.2}$$

$$I_2 = \int_0^{\pi}\int_0^{\pi}\left[\frac{\sin(\pi W / \lambda \cos\theta)}{\cos\theta}\right]^2 \sin^3\theta \cdot\cos^2(\pi L_e / \lambda \sin\theta \sin\varphi)d\theta d\varphi \tag{5.3}$$

where W and L_e are the width and effective length of patch, and λ is the free-space wavelength at resonance. The directivity D_t is a function of the electrical size of patch with respect to wavelength, that is, W/λ and L_e/λ.

As these pins are introduced, the resonant frequency of the MPA tends to shift from f_0 to f_r, while its free-space wavelength at resonance varies from λ_0 to λ. Taking the fringing effect into account, the effective length L_e of the patch is a function of its width W, its length L, substrate thickness h, and relative permittivity ε_r. The radiation directivity D_t is dependent on all these involved parameters, f_r/f_0, W/L, h/λ_0, and ε_r, in such a general function that

$$D_t = F'(\frac{f_r}{f_0}, \frac{W}{L}, \frac{h}{\lambda_0}, \varepsilon_r) \tag{5.4}$$

According to [2], the thickness of the substrate has minimal impact on directivity of the thin antenna. When the substrate and the size of the MPA are fixed, the enhanced directivity is primarily influenced by the frequency tuning ratio f_r/f_0. Radiation directivity of the proposed square patch antenna is examined with a fixed patch size at first. The radius of the pins is set as $R = 0.5$ mm. Initially, a square patch antenna with a fixed length and width ($W = 63.0$ mm) is considered. The parametric study in Figure 5.2 illustrates the resonant frequency and directivity as a function of the pin-to-pin spacing D/W. When all four shorting pins are positioned at the center of the patch (i.e., $D/W = 0$), the directivity and resonant frequency are 7.5 dBi and 1.50 GHz, respectively, which are identical to those of a conventional square patch antenna without shorting pins. As D/W increases from 0 to 1, both directivity and resonant frequency increase to their peak value at about $D/W = 0.71$, and then decrease. At $D/W = 0.71$, the maximum resonant frequency and directivity are approximately 2.18 GHz and 9.4 dBi, respectively, indicating an increase in radiation directivity at resonance by about 1.9 dB. If the resonant frequency of the radiator is fixed, the antenna's directivity and electrical size (W/λ) are enhanced as a function of pin location D/W, also shown in Figure 5.2.

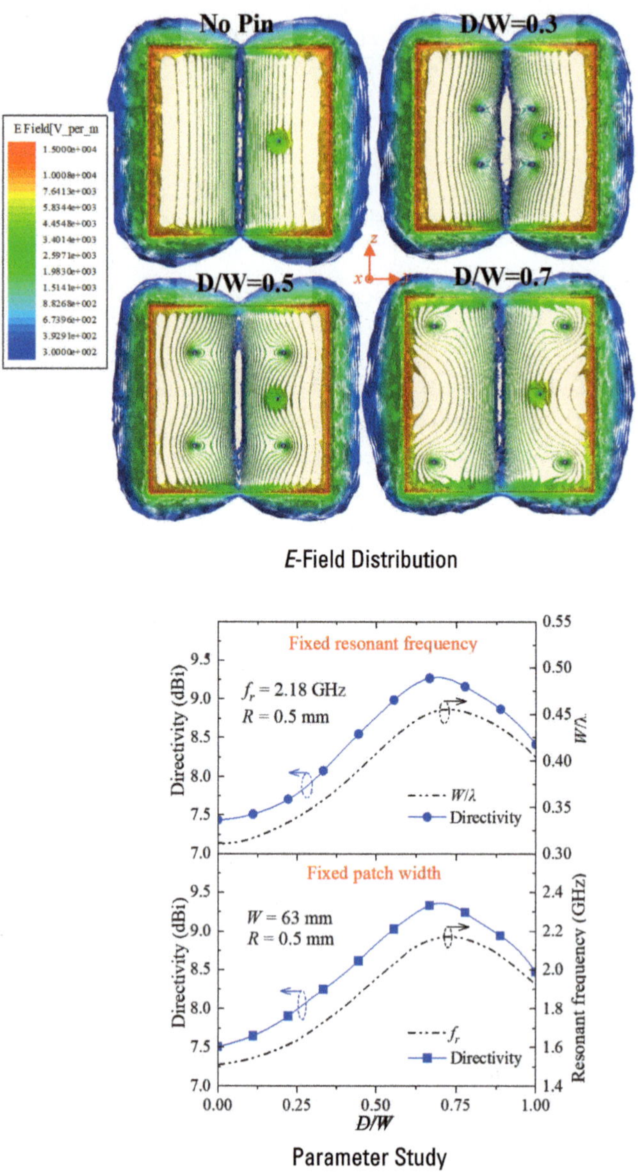

Figure 5.2 Electric-field distribution beneath the patch under different pin-to-pin spacing. Directivity, resonant frequency, and normalized width of patch versus pin-to-pin spacing.

To explain the underlying physics, the *E*-field distribution in Figure 5.2 discusses the electric-field distributions of the proposed MPA and quantitatively shows its resonant characteristics. For a square MPA without any pins under TM_{01} mode resonance, the electric-field intensity beneath the patch can be intuitively considered to vary as a cosine function along the *y*-axis and be constant along the z-axis. However, once the four pins with a radius of 0.5 mm are introduced, the field distribution is perturbed due to shunt inductive effects caused by the pins. The figures illustrate the field distribution under different pin-to-pin spacings, with D/W values of 0.3, 0.5, and 0.7. As the four pins are moved outward from the center toward the four edges of the square patch, the electric field beneath the patch is increasingly perturbed, resulting in a boost in its resonant frequency.

To validate these principles, two square patch antennas with and without four shorting pins were designed and fabricated on a Rogers 5880 substrate with $\varepsilon_r = 2.2$, $h = 3.175$ mm, and 2.8 μm surface roughness of copper. A photograph of the fabricated patch antenna with four shorting pins, and a conventional patch antenna with the same size, are illustrated in Figure 5.3. In the final design, the radius and locations of pins have been optimized until the directivity of the antenna reaches its maximum value, and the pin configuration is finally determined to be $D/W = 0.6$, $R = 4$ mm.

Figure 5.3 also compares the simulated and measured radiation patterns between these two types of antennas. The patch antenna with four shorting

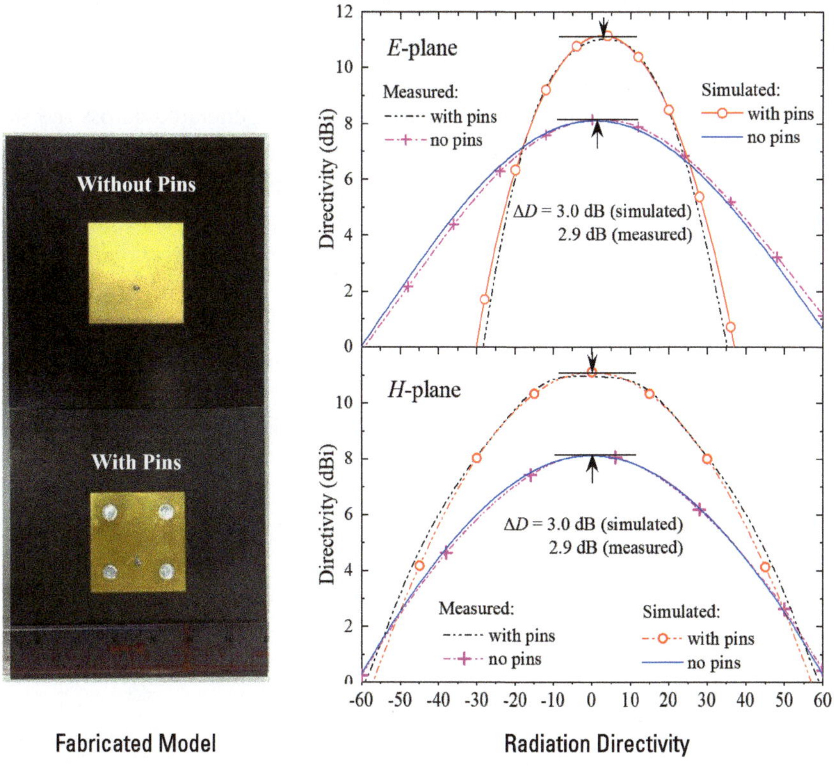

Figure 5.3 Photographs of the patch antennas with and without shorting pins under the same physical size ($D/W = 0.6$, $R = 4$ mm). Radiation directivity is also compared.

pins has a much narrower HPBW compared to its counterpart without pins. In particular, the *E*-plane HPBW is reduced by 32°. As a sequence, the proposed antenna gets a directivity enhancement of about 2.9 dB. The results in simulation and measurement also indicate that the two antennas exhibit high-radiation efficiency.

Shorting pins are loaded underneath the radiating patch to increase its resonant frequency and electrical size. As a result, the gain of the rectangular patch antenna is successfully enhanced under fundamental mode resonance.

5.2.2 Slots and Shorting Pins for Fundamental Mode

Although the technique in Section 5.2.1 effectively enhances gain, the electrical size of the patch is enlarged simultaneously. In this section, a patch antenna with loading of slot and shorting pins is proposed to enhance radiation gain without increased aperture size [4]. The working principle of a proposed directivity-enhanced patch antenna will be discussed and demonstrated through the theoretical analysis under an equivalent array model.

Figure 5.4 illustrates a diagram of the proposed gain-enhanced antenna without an increased patch size. In this configuration, a transverse slot is positioned along the central line of a patch, where the electric field of the fundamental TM_{01} mode is null underneath the patch, and its surface current reaches a maximum. The *y*-oriented tangent electric field is generated across the slot, making this transverse slot radiative. The antenna can be viewed as three radiation slots with equivalent magnetic currents, namely two M_{S0} and one M_{S1}, as shown in Figure 5.4. The tangent field in the central slot is in the opposite direction, or out of phase, compared to those in the other two slots. As shown in the equivalent magnetic current model, the antenna can be simplified as a three-element array, and the total electric field at a far-zone observation point, *P*, is attributed to these three magnetic currents. Finally, the radiation patterns of the antenna with and without the slot are compared.

Based on the previous discussion, the *E*-plane far-field radiated by the two slots at two sides can be derived as below [3]

$$\begin{aligned} E_{q01} &= C_{01}\,\mathrm{sinc}(h\sin\theta/\lambda)\cos(\pi W\sin\theta/\lambda) \\ &\approx C_{01}\cos(\pi W\sin\theta/\lambda) \end{aligned} \tag{5.5}$$

where h is the thickness of antenna, W the width of square patch, λ the free-space wavelength. The coefficient C_{01} is given by

$$C_{01} = j\frac{2WV_{01}e^{-jkr}}{\lambda r} \tag{5.6}$$

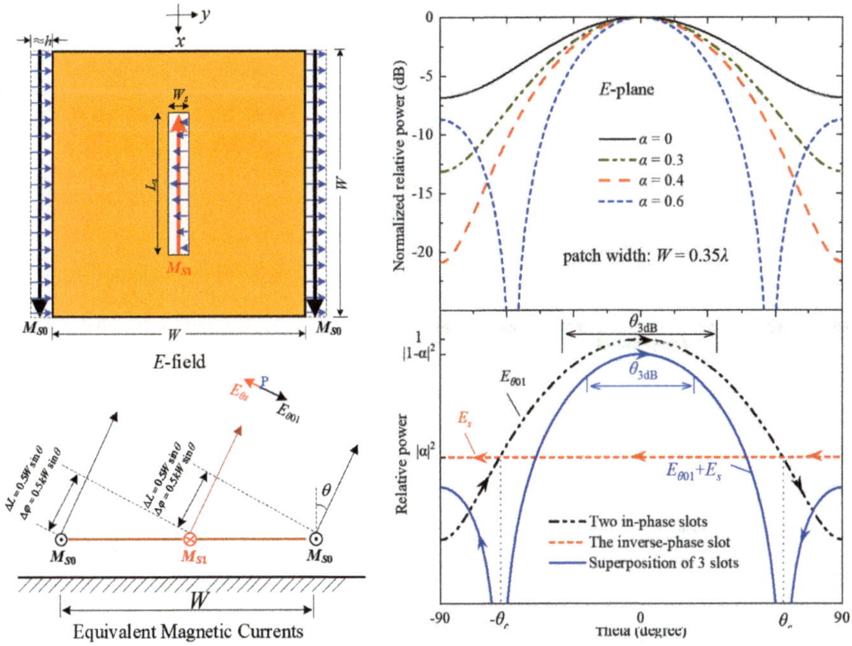

Proposed Patch

Radiation Pattern Variation

Figure 5.4 Diagrams of the proposed gain-enhanced patch antenna, and varied radiation patterns comparison.

where k, V_{01}, and r are the propagation constant, the voltage across the slot, and the distance from the far-zone point to the center of the patch.

The electric-field intensity in the central slot should be modeled as a cosine distribution with respect to its length due to the two shorted ends of this slot, and the E-plane far-field radiated by it can be derived and simplified as [3]

$$E_{\theta s} = -C_s \operatorname{sinc}(W_s \sin\theta / \lambda) \approx -C_s \tag{5.7}$$

where the coefficient C_s is related to the length of slot, that is, L_s, and amplitude of slot voltage in the center, that is, V_s, and it can be eventually expressed as

$$C_s = j\frac{2L_s V_s e^{-jkr}}{\lambda \pi r} \tag{5.8}$$

Thereafter, the total far field from this entire patch antenna can be calculated with linear superposition of $E_{\theta 01}$ and $E_{\theta s}$, such that

$$E_{\theta T} = E_{\theta 01} + E_{\theta s} = C_{01}\left[\cos(\pi W \sin\theta / \lambda) - \alpha\right] \quad (5.9)$$

$$\alpha = C_s / C_{01} \quad (5.10)$$

where α represents the amplitude ratio of far fields radiated by the central slot and the other two slots.

As shown in Figure 5.4, the power radiated by two in-phase slots is a cosine square function of $\pi W \sin\theta/\lambda$ at *E*-plane, and relative power in broadside is set as 1 for reference. The relative power radiated by the out-of-phase central slot is a constant of $|\alpha|^2$ over the angular range. If the electrical width of the patch is kept as a constant after the slot is loaded, the far field is determined by the linear superposition of $E_{\theta 01}$ and $E_{\theta s}$. Since $E_{\theta 01}$ and $E_{\theta s}$ are out of phase, they will cancel each other at specific angles of $\pm\theta_c$, where their intensities are exactly the same. The total field becomes zero at $\pm\theta_c$ and the sidelobes come up. The HPBW is thus significantly narrowed, and the antenna becomes more directive.

The *E*-plane radiation patterns of a patch antenna with a width of 0.35λ are computed for various α ratios, and the results illustrated. When $\alpha = 0$, the radiation pattern is identical to that of a conventional patch antenna without any slot, displaying a wide radiation beamwidth. As the field intensity radiated by the central slot gradually increases, or α increases from 0 to 0.3 and 0.4, the radiated power near the grazing is diminished. When α reaches 0.6, radiation nulls appear at $\theta = \pm 57°$ and the sidelobes become more pronounced. The HPBW becomes increasingly narrow, and radiation directivity is progressively enhanced.

Next, simulation is used to verify the proposed principle. Figure 5.5 shows the suspended configuration of the patch antenna, with substrate thickness and suspended height set at $h_1 = 1.57$ mm and $h_2 = 1.57$ mm, respectively. The square patch is positioned on the *x*-*y* plane with its center at (0, 0) and a width of $W = 30$ mm. A transverse narrow slot with a width of $W_s = 2$ mm and a length of L_s is longitudinally located in the *y*-direction. Additionally, two shorting pins with a radius of *R* are longitudinally placed at the two symmetrical sides of the central slot in the *y*-axis with a spacing of *d*. In this study, a Rogers 5880 substrate with a relative permittivity of $\varepsilon_r = 2.2$ is used.

To be extended, Figure 5.6 evaluates the radiation performance of the proposed antenna. Directivity is progressively enhanced to about 10.0 dBi as the slot length increases, especially in the range of L_s/W from 0.5 to 0.9. This means 1.6-dB increment in radiation directivity compared to a conventional MPA with $L_s/W = 0$. Patch size and resonant frequency are both kept unchanged by adjusting the pin spacing. As directivity is enhanced, radiation efficiency is reduced, since the far-field cancellation caused by the three slots increases the *Q*

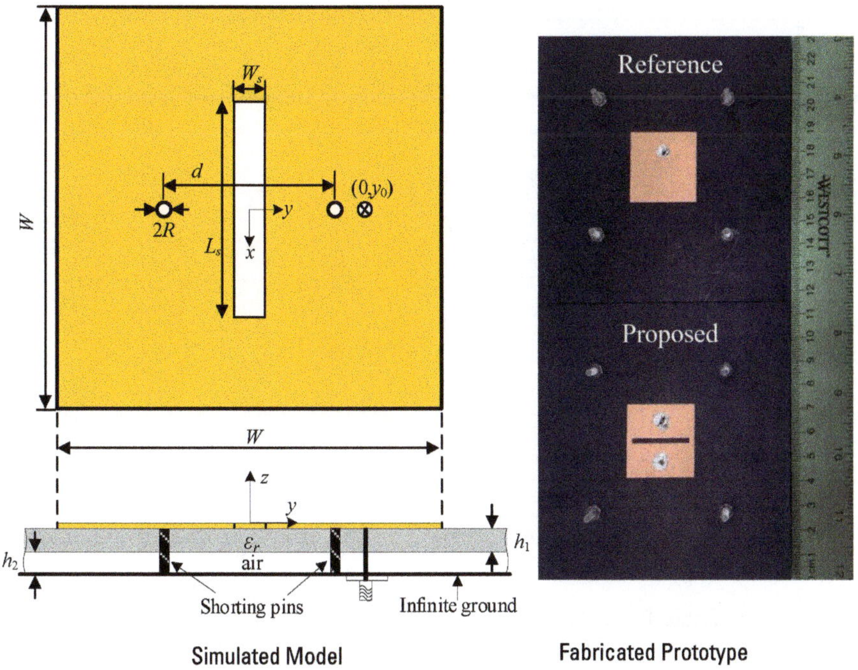

Figure 5.5 Geometry and a fabricated prototype of the patch antenna loaded with a transverse slot and a pair of shorting pins along the longitudinal central line of a patch.

factor of the antenna, increasing the conductor and dielectric losses. As shown in Figure 5.6, radiation gain first rises, then begins to decline after it reaches its maximum value. The maximum radiation gains under slot widths of 1 and 2 mm are equal to 9.5 and 9.3 dBi with the corresponding value of L_s/W to be 0.87 and 0.85, respectively.

To experimentally validate the proposed principle, two antenna prototypes were designed, fabricated, and tested. Photographs with and without loading of slot/pins are provided in Figure 5.5. In the implementation, the substrate is suspended above the ground at a height of 1.57 mm. Both patches have a width of W = 30 mm, and the ground plane size is 115 mm × 115 mm. The feeding points of the proposed and conventional patch antennas are set at y_0 = 6.8 and 5.0 mm, respectively. In the proposed patch antenna, the slot is specified with W_s = 2 mm and L_s/W = 0.85, while the shorting pins are with R = 0.4 mm and d/ W = 0.57.

Simulated and measured radiation patterns are also plotted in Figure 5.6. The antenna significantly narrows the main beam with radiation nulls caused by linear superposition from the three slots. In particular, the HPBW in the E-plane is reduced by 22° and 20° in simulation and measurement, respectively.

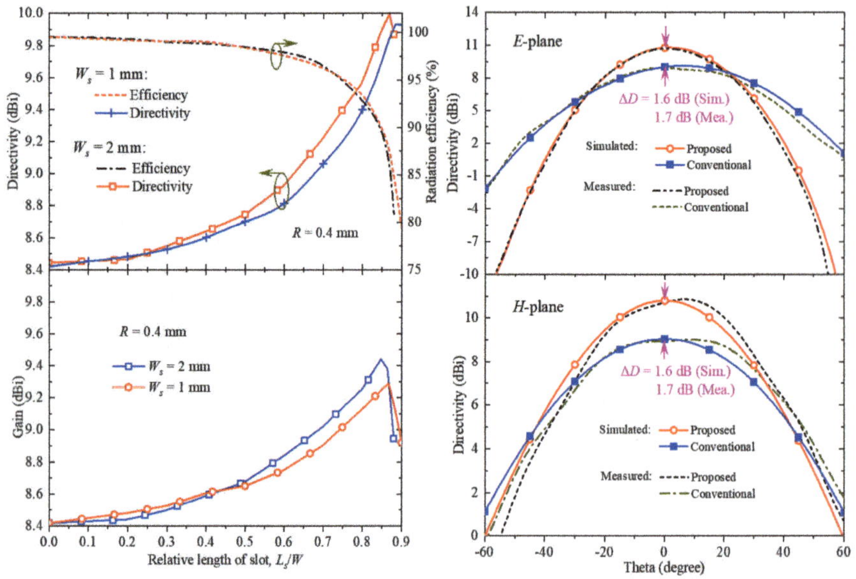

Radiation Performance of the Proposed Patch

Radiation Pattern Comparison

Figure 5.6 Radiation performance of the patch antenna in Figure 5.5 under different slot size at fixed 3.58 GHz. Radiation patterns of the proposed antenna and conventional antenna at the resonant frequency are also compared.

The simulated and measured directivities of the proposed antenna are enhanced to about 10.8 and 10.7 dBi, with an increment of 1.6 and 1.7 dB, respectively, compared to its conventional counterpart.

In summary, the slot is cut at the center of patch to generate an additional out-of-phase radiation source, and shorting pins help keep constant resonant frequency. The far-field cancellation effect of the three radiating slots finally contributes to a narrower HPBW in the *E*-plane. The gain of the antenna is successfully enhanced without increased patch size at the fundamental TM_{01} mode.

5.2.3 Slots and Shorting Pins for High-Order Modes

Sections 5.2.1 and 5.2.2 discuss how to enhance the radiation gain of a patch antenna operating in single mode and a single band. In this section, a patch antenna with dual-band high gain is further proposed and studied. The primary working principle is to enhance the gain of the fundamental mode with shorting pins and suppress the sidelobe of high-order mode with slot loading. For more details, readers can refer to our reported work [5].

Figure 5.7 plots the configuration of the proposed antenna, which consists of a radiating patch, shorting pins, slot, and differential feeding. The substrate is F4B, with relative permittivity of 2.2, thickness of 2 mm, loss tangent of 0.0007, and copper surface roughness of 2.8 μm. TM_{11} and TM_{12} modes are excited for the circular patch antenna. TM_{11} mode operates in the lower band, and TM_{12} mode operates in the upper band.

The effect of shorting pins on the lower band is investigated in Figure 5.8. As shown in the results of the pin-loaded TM_{11}-mode antenna, when R_s/R_p is varied from 0 to 0.8, resonant frequency is increased from 2.454 to 4.116 GHz. This means that the electrical diameter of patch at resonance is enlarged on a scale of 1.68. The directivity at resonance is enhanced from 7.4 to 10.6 dBi with 3.2-dB increment. Meanwhile, the HPBWs in the *E*- and *H*-plane are substantially narrowed from 98.5° and 80.5° to 53.5° and 66.2°, respectively. In the range of 3.65 to 3.70 GHz, the directivity of the antenna is around 10.0 dBi.

Next, the effect of the slot length L_s on the resonant frequencies of its TM_{11} and TM_{12} modes is further discussed as shown in Figure 5.8. The

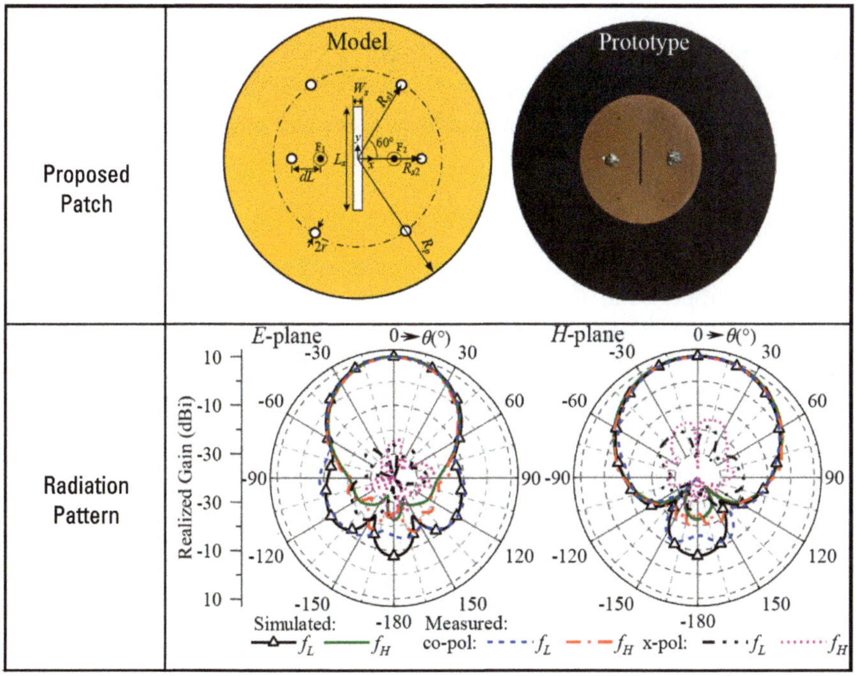

Figure 5.7 Geometry and prototype of the proposed differentially fed circular patch antenna with loading of six pins and a central slot. Radiation patterns are also plotted. Geometric parameters of the antenna: $r = 0.6$ mm, $W_s = 1$ mm, $R_p = 23$ mm, $L_s/R_p = 0.8$, $R_s/R_p = R_{s1}/R_p = 0.72$, $R_{s2}/R_p = 0.60$.

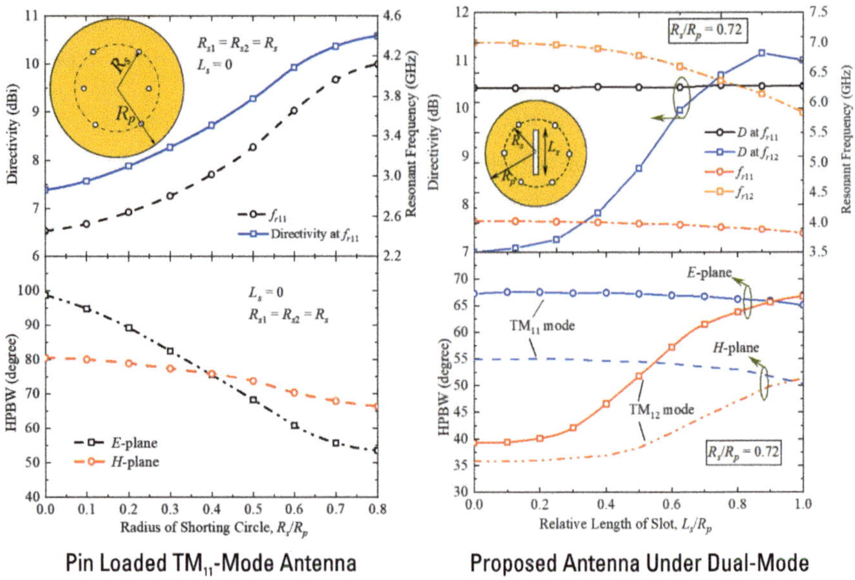

Figure 5.8 Simulated results of the proposed patch antenna under different modes with different loading schemes.

resonant frequency, directivity, and HPBWs in lower band are all stable when L_s is increased from 0 to R_p because L_s is sufficiently short with respect to the wavelength and patch diameter. In contrast, the radiation performance in upper band is more sensitive. The resonant frequency of TM_{12} mode is reduced from 6.991 to 5.388 GHz as L_s/R_p ranges from 0 to 1, resulting in the f_{r12}/f_{r11} declining from 1.739 to 1.468. Directivity is enhanced from 7.0 dBi to about 11.1 dBi, then slightly rolled down. For beamwidths, the *E*-plane and *H*-plane HPBWs in upper band are broadened as L_s/R_p is increased. As L_s/R_p is over 0.8, the HPBWs in two bands become very close, indicating the attractive features of almost the same beam in the two bands. For this patch, shorting pins are placed at $R_s/R_p = 0.72$, which is near the electric-field null of the TM_{12} mode, in order not to perturb this high-order mode. The slot helps reshape the current distribution of the TM_{12} mode, and radiation gain is enhanced because the *E*-plane sidelobes are well suppressed by the slot.

To validate the anticipated performance, an antenna prototype was manufactured and tested. The simulated and measured differential reflection coefficients, $|S_{dd}|$, of the antenna are also shown in Figure 5.9. The actual test results of $|S_{dd}|$ are below –10 dB in dual band, such as from 3.654 to 3.712 GHz, and from 5.844 to 5.976 GHz.

The realized gain and efficiency in both simulation and measurement are illustrated in Figure 5.9. The maximum simulated and measured efficiencies of the antenna in the dual bands exceed 90%. Measured gains at the two central

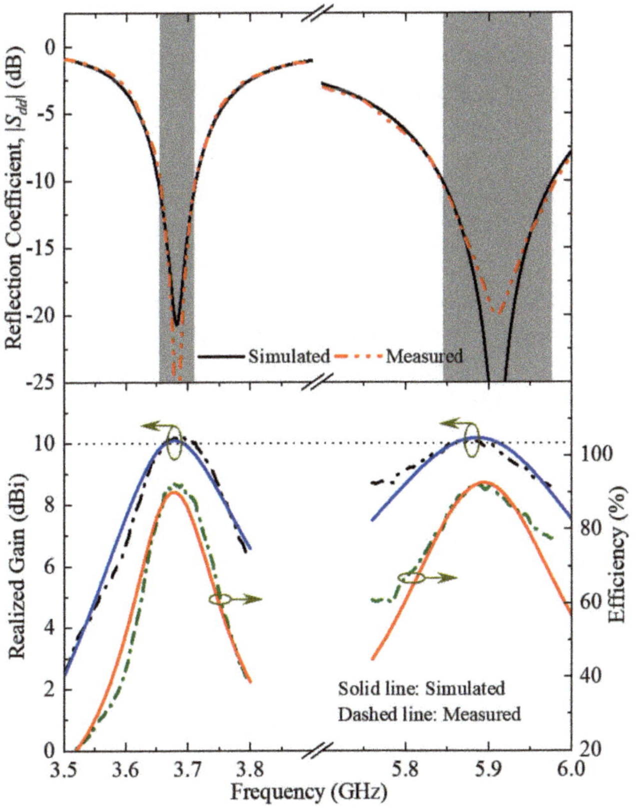

Figure 5.9 Measured S-parameter, gain, and efficiency of the prototype.

frequencies of 3.675 and 5.888 GHz are as high as 10.2 and 10.0 dBi, respectively, significantly surpassing those of conventional dual-band patch antennas.

Additionally, dual-frequency radiation patterns in the two principal planes are shown in Figure 5.7. The antenna at the two frequencies exhibits main beams with nearly identical beamwidths in both the *E*- and *H*-planes, despite the substantial frequency ratio of the two bands. On closer examination of these two bands, the backside lobes at 3.675 GHz are relatively higher than those at 5.888 GHz, but the front-to-back ratio remains above 17.7 dB, which is still acceptable in many applications.

In summary, shorting pins are loaded to increase the electrical size of patch and enhance the radiation gain of the antenna in the lower band under TM_{11} mode resonance. Meanwhile, the linear slot is cut on the radiating patch to reshape the current distribution of high-order TM_{12} mode and optimize the gain in the upper band. Antenna gain is successfully enhanced under TM_{11} and TM_{12} modes. Most important, stable HPBWs are maintained in the two bands, especially with a large frequency ratio.

Various methods for enhancing high-gain bandwidth are shown in Figure 5.10. In [6], a single nonresonant narrow slot introduced at the center of the patch reduces the *E*-plane sidelobe of the TM_{12} mode of the circular microstrip patch. This reduction is due to the out-of-phase field superposition between the patch and the slot at the sidelobe positions, resulting in high gain. In another design [7], wide bandwidth is achieved by merging the TM_{01} mode generated by the shorting vias on the circular patch, the TM_{02} mode of the patch, and the TM_{02} mode of the annular ring. Similarly, in [8], varying the radii of the vias on the circular patch antenna enables additional excitation of the TM_{11} and TM_{13} modes, leading to wideband or dual-band performance with high-gain and sidelobe suppression. The radiation fields of these modes superpose with the radiation fields of the antenna's TM_{12} and TM_{14} modes. In [9], a dual-compressed mode dual-polarized patch antenna is proposed. By introducing slots and vias on the patch, the TM_{03} mode is compressed to achieve high directivity while the TM_{05} mode is made to move closer to the TM_{03} mode, enhancing the operating bandwidth. There is also a design whereby selecting an appropriate value of the dielectric constant, both high directivity and low-sidelobe level (SLL) can be achieved for a disc antenna operating in the TM_{12} mode [10].

5.3 Stable Gain Bandwidth Technique

From the above section, although the gain of the MPAs is obviously enhanced, they still suffer from narrow bandwidth in each operating band. Practical antennas should have wide bandwidth and stable radiation gain. This section discusses the working principle for realizing stable high gain and widened bandwidth simultaneously. In Section 5.3.1, the gain bandwidth of an antenna is extended by constructing a third-order filter network through additional feeding resonators. In Section 5.3.2, the multimode technique is employed, and the TM_{31} and TM_{12} modes of a circular patch antenna are simultaneously excited to achieve wideband and stable high gain.

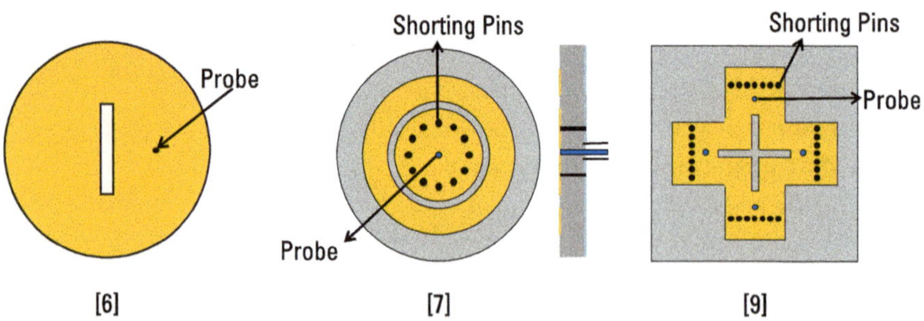

Figure 5.10 Geometry of antennas in the references.

5.3.1 Wideband Stable Gain via Feeding Resonators

A resonator-fed patch antenna is proposed for wideband high gain, improved selectivity, and reduced cross polarization. The patch antenna design takes inspiration from the cascaded trisection (CT) filter [11]. Illustrated in Figure 5.10 are the exploded, top, and side views of the antenna. The antenna features two substrates, upper and lower, with thicknesses identified as T_1 and T_2, respectively. The upper substrate is home to a square patch while the lower substrate carries two resonators. A meandered, spider-shaped resonator excited by a vertical coaxial probe is nested within a rectangular ring resonator. The two substrates are Rogers 4003C, with relative permittivity of $\varepsilon_r = 3.55$, thickness of $T_1 = 0.81$ mm, and $T_2 = 0.508$ mm.

To achieve sufficiently high gain, a square patch antenna working in TM_{03} mode is employed, with the working principle explained in [6, 7]. The patch geometry is shown in the inset of Figure 5.12. Input impedance and mode distribution of a slot-loaded square patch antenna under higher order mode operation. Adjacent modes, such as TM_{02}, TM_{20}, and TM_{22}, are even modes. In our proposed design, both feeding resonators operate in half-wave resonance, of which the field and current distribution are of odd symmetry. The feeding resonators can suppress the undesired even modes and enhance polarization purity. The design is carried out with the following steps.

First, the slot-loaded and current-reshaping technique in [12] is further developed in this work for sidelobe suppression and gain enhancement. Slot length (L_s) determines the E-plane sidelobe level, while slot spacing (D) affects the H-plane beamwidth. These slot parameters were set as $L_s/W = 0.35$ and $D = 15.4$ mm by considering both the achievable gain and the radiator's quality factor.

Following this, the antenna is equivalent as a third-order CT filter, which will widen the bandwidth and improve frequency selectivity. The equivalent circuit model is shown in Figure 5.11, where the radiator is equivalent as the last-stage lossy resonator. To be simplified, the network is assumed to be symmetric, that is, $f_{01} = f_{03}$, $M_{12} = M_{23}$, $Q_{e1} = Q_{e2}$. Through a weak-coupling extraction, the quality factor of the patch radiator is $Q_{e2} = 24.5$. After that, the in-band reflection coefficient is selected to be below -11.5 dB, and the attenuation pole is set at $f = 3.75$ GHz. The coupling coefficients (M_{12}, M_{23}, and M_{13}) and resonant frequencies (f_{01}, f_{02}, and f_{03}) can be derived using the method outlined in [13]. The values derived are presented in Figure 5.10. With these parameters, the reflection coefficient of the antenna can be determined and predicted in a closed-form manner. In our reported work [11], we provide a detailed calculation process. For brevity, it is not shown here.

To explain the working principle, Figure 5.13 compares the on-patch current distribution of our proposed antenna with that of a reference antenna. The reference antenna loses the odd symmetry of current distribution with

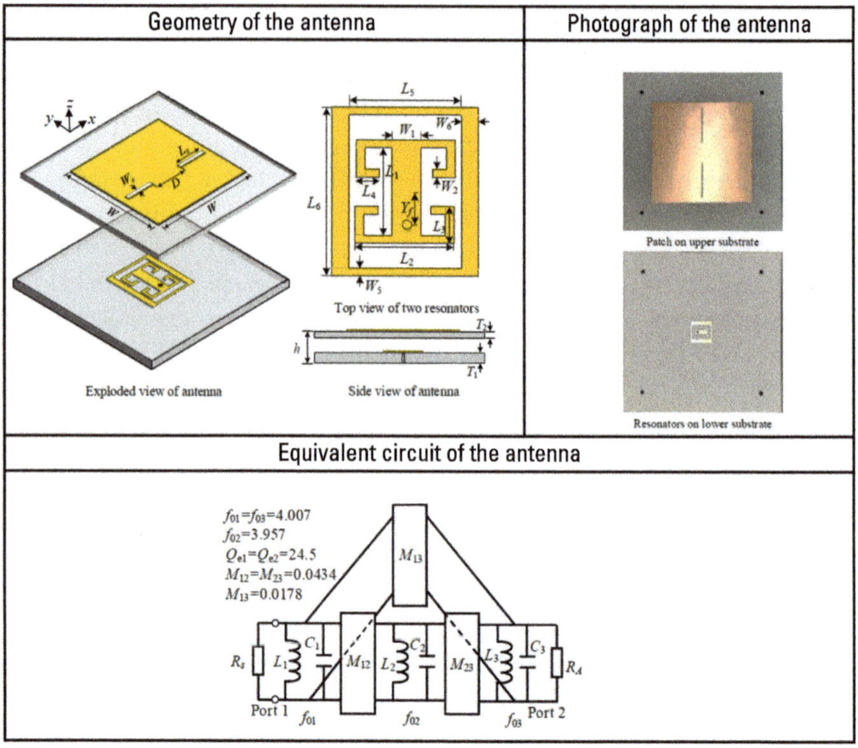

Figure 5.11 Geometry, photograph, and equivalent circuit of the proposed patch antenna.

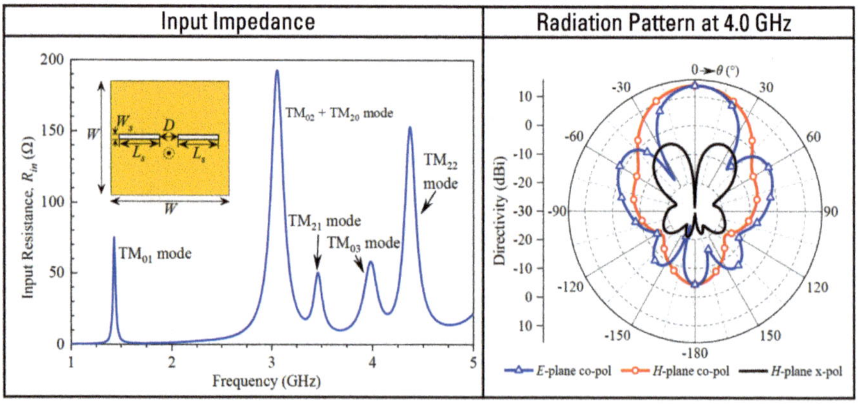

Figure 5.12 Input impedance and mode distribution of a slot-loaded square patch antenna under higher order mode operation. The patch is directly fed by a probe in this case.

respect to *H*-plane, due to the existence of other even modes, which arouses the cross polarization. In contrast, the current distribution of the proposed patch

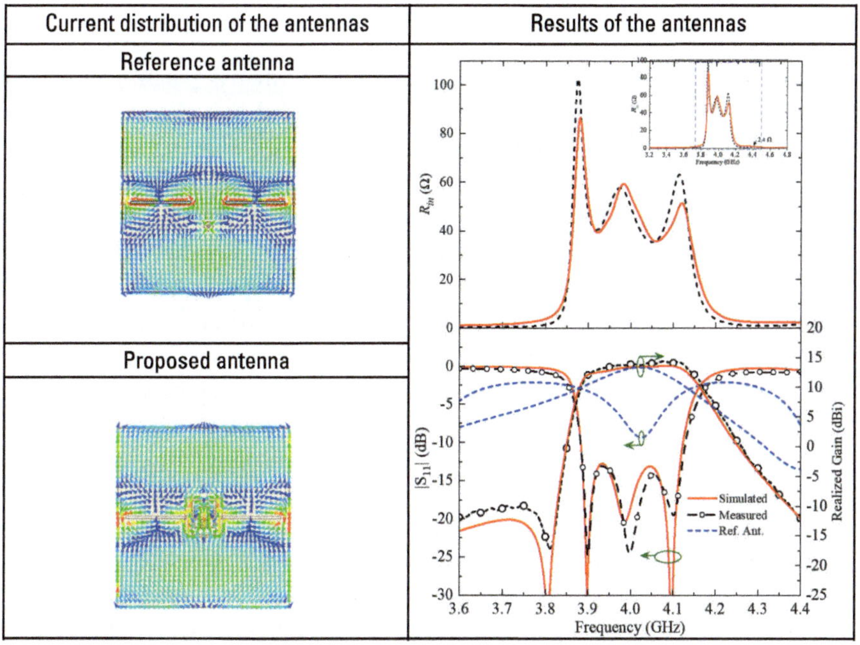

Figure 5.13 On-patch current distribution at 4 GHz, and simulated and measured results of the reference and proposed antennas.

antenna is more symmetric, as adjacent even modes have been well suppressed. This improved current symmetry results in reduced cross polarization.

An antenna prototype was fabricated to verify the principle; the antenna photograph is shown in Figure 5.11. The two substrates are assembled with plastic screws, and a coaxial probe is soldered on the bottom side of the lower substrate. The reflection coefficient and realized gain of the proposed antenna and reference antenna are shown in Figure 5.13. The reference antenna experiences a gradual reduction in realized gain at frequencies away from the central frequency of 4 GHz, resulting in poor selectivity due to its narrow bandwidth and poor matching. In contrast, the bandwidth of the proposed antenna is much wider. As expected, three reflection zeros have been produced, and the $|S_{11}|$ below –10 dB ranges from 3.882 (3.884) to 4.130 (4.123) GHz in measurement (simulation). The proposed antenna achieves a flat realized gain response in the passband, with a maximum simulated gain of 13.6 dBi and a measured gain of 14.4 dBi. This is because introducing two resonators below the radiating patch produced two additional resonances for wider bandwidth while the TM_{03} mode dominates the high-gain radiation. Compared to previous works, the proposed multimode resonance-inspired antenna achieved stable high gain in band, enhanced selectivity out of band, and produced a more compact structure by constructing a third-order CT filter topology with feeding resonators.

5.3.2 Wideband Stable Gain Via Dual Modes

In this section, the TM_{31} and TM_{12} modes of a circular patch antenna are simultaneously used for stable gain over the wideband [14]. Figure 5.14 plots the configuration of the proposed antenna loaded with pins and slots, where a circular patch is suspended above the ground with a height of H, and is loaded with three cascaded slots and two shorting pins. The slot loading in this design can significantly reshape the radiation patterns of the two operating modes. Two open slots are used to reshape the current of TM_{31} mode, and the central slot is used to manipulate the TM_{12} mode. Two symmetrical shorting pins with a radius of R_p are placed on either side of the slots, with a spacing of D_p to further adjust the resonant frequencies of these two modes to form wide operating bandwidth. To enhance impedance matching, two feeding probes with caps on the top are coupled to the patch using a differential excitation that utilizes capacitive feeding [15].

As illustrated in Figure 5.15, when the TM_{31}-mode patch has no slot loading, its on-patch current is strong outside and weak inside, resulting in zero density at the patch center. The current vectors of every two adjacent semiperiods are out of phase, causing them to cancel each other in the broadside direction and resulting in a conical radiation pattern. This cancellation also contributes to a relatively high-quality factor of TM_{31} mode. When two open slots are loaded onto the patch edge, the current vectors of two semiperiods begin to circulate around the slots. As slot length increases, these current vectors become more tightly coupled to the sides of the slots. Because of their close spacing and symmetry, they cancel each other and produce little far-field radiation. Only the four remaining semiperiods of currents, with effective components in the y-direction, dominate the radiation. The maximum radiation intensity is in the broadside direction in this scenario.

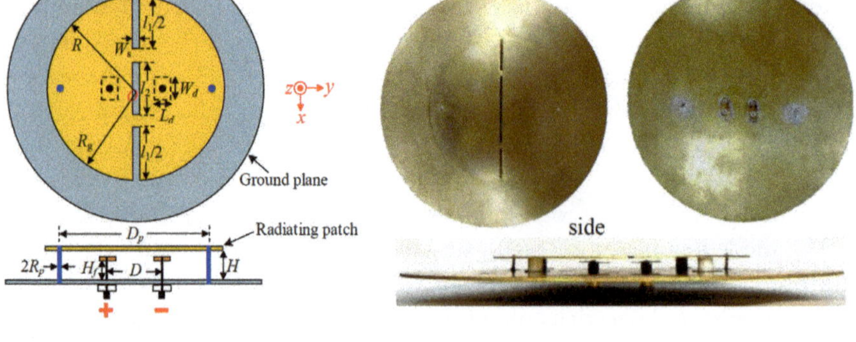

Figure 5.14 Geometrical configuration and photograph of the proposed circular MPA.

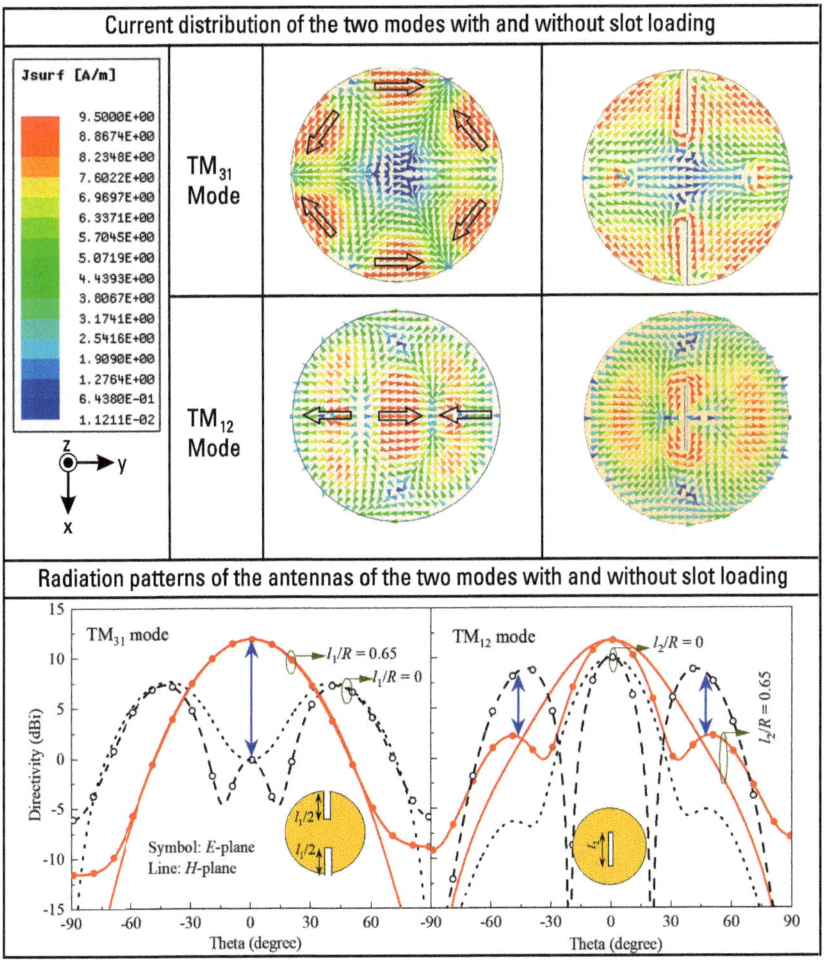

Figure 5.15 On-patch current distribution of the two modes with and without slot loading.

Figure 5.15 also shows the current distribution of the TM_{12} mode on the patch. Without slot loading, the density of the current distribution is strong on the inside and weak on the outside, differing from that of the TM_{31} mode. The y-oriented current vectors along the radial direction resemble those of a 3-$\lambda/2$ dipole, with the central part being out of phase, resulting in high SLL of the antenna. After the central slot is introduced, the out-of-phase current is forced to flow around the slot. As slot length increases, the interrupted current becomes more tightly coupled to the slot, leading to a reduction in effective radiation. By properly selecting slot length and changing current distribution, the sidelobe radiation can be well suppressed.

Next, the effect of the slot on the radiation patterns of two modes is investigated in Figure 5.15. For the TM_{31}-mode patch without a slot, the maximum radiation direction is around $\theta = 45°$ and the broadside intensity is significantly lower. With the introduction of two open slots, the pattern shape is transformed from conical to broadside, and the broadside directivity is increased to a high level due to the reshaped current distribution. For the TM_{12} mode, the *E*-plane SLL normalized with broadside directivity is about –1.1 dB in the case without slot loading. After introducing a slot with a length l_2/R, the SLL is reduced to –9.6 dB. Broadside directivity is enhanced from 10.0 to 11.8 dBi, while the *H*-plane HPBW is slightly widened.

Antenna impedance performance is improved through the use of hybrid loading, which involves investigating three different loading schemes. The analysis in [14] demonstrates that the first scheme's two-sided open slots only affect the lower band as contributed by TM_{31} mode, whereas the second scheme's central slot only affects the upper band related to TM_{12} mode. This means that each set of slots can independently adjust the resonant frequencies of the two modes. Shorting pins are located where the TM_{12}-mode field is weak and TM_{31}-mode field is strong, resulting in an upward shift of the TM_{31} resonant frequency. This produces a continuous and wide operating band with $|S_{dd}|$ below-10 dB. For more details about the impedance, readers can refer to our published work in [14].

The antenna's *E*- and *H*-plane HPBWs show narrow and extremely steady beamwidths. In the frequency range of 3.3 to 3.7 GHz, the *E*-plane HPBW remains almost constant between 30.2° and 31.8°, while the H-plane HPBW varies slightly from 39.8° to 44.2°. Associated figures can be found in [14]. The consistent and narrow beam is primarily due to reshaped current distribution. To confirm the radiation improvement resulting from the two modes, 3D radiation patterns for TM_{31} mode at 3.35 GHz, and TM_{12} mode at 3.62 GHz, are carried out and shown in the inset of Figure 5.16. Thanks to the current reshaping effect brought about by slot loadings, the two modes exhibit very consistent radiation beams, which is highly desirable for wideband antennas.

An antenna prototype was designed, fabricated, and measured. The prototype is shown in Figure 5.13, and the measurement results of the antenna are shown in Figure 5.16. The patch and ground are made of 0.5-mm-thick copper plate. This design produced two reflection zeros corresponding to the TM_{31} and TM_{12} modes, resulting in a broad operating bandwidth of approximately 12.4% (3.295 to 3.730 GHz), while the measured one is 13.4% (3.265 to 3.735 GHz). The antenna's realized gain within this band is consistent with very little variation, reaching a high level of 13.7 dBi.

During the passband of 3.3 to 3.7 GHz, the beamwidths in both *E*- and *H*-planes are relatively narrow, and the main beams are kept nearly unchanged, leading to stable high gain. Sidelobes in the *E*-plane are still present but are effectively suppressed below –16 dB. The symmetric feeding and antenna geometry also contribute to well-suppressed cross polarization.

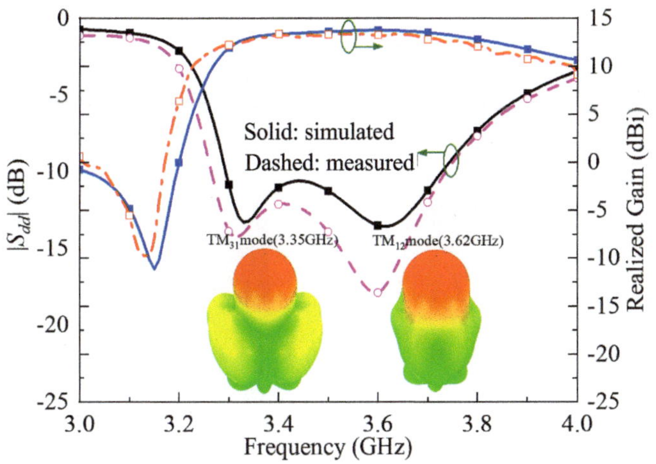

Figure 5.16 Realized gain, and differential reflection coefficient of the fabricated antenna prototype.

In summary, the current reshaping technique via slots is proposed to simultaneously manipulate the radiation patterns of two high-order modes as desired. Meanwhile, the shorting pins are loaded to transform the conical-like radiation patterns of TM_{31} mode into broadside main beam with high gain. Additionally, the sidelobes of TM_{12} modes are significantly reduced. Finally, comparing with these previous works, the proposed multimode resonance inspired antenna has achieved stable high gain, wide bandwidth, and low cross polarization.

Other designs aimed at achieving stable high gain are shown in Figure 5.17. For instance, in [16], a microstrip patch antenna with both sides shorted is proposed, where the TM_{11} and TM_{03} modes are excited and combined to achieve a wide bandwidth and high gain. This is achieved through two long radiation apertures functioning as cosine magnetic currents. In another work, [17] introduces a compact differential fed dual-polarized filtering patch antenna featuring DGS, a cross slot, and eight shorting pins. The DGS provides suppression in the lower stopband while the cross slot offers suppression in the upper stopband. The shorting pins are used to adjust the patch resonance, enabling a broader bandwidth. In [18], a rectangular ring patch operating in the TM_{12} mode is embedded with a rectangular patch containing six shorting vias, which works in the TM_{10} mode. The reshaped current distribution of the TM_{12} mode enhances boresight gain and generates a radiation null at the lower band edge. Meanwhile, the embedded rectangular patch introduces another radiation null at the upper-band edge, and this integrated structure effectively broadens the passband bandwidth.

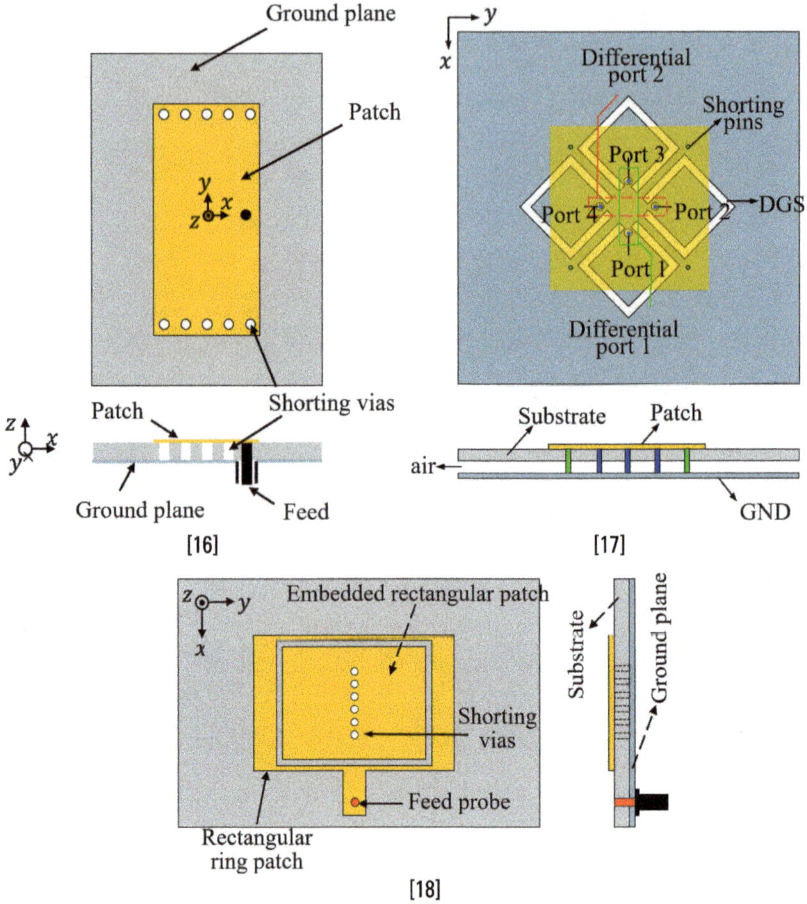

Figure 5.17 Geometry of antennas in the references.

5.4 AR Bandwidth Enhancement Technique

Previous sections discuss LP performance. However, applications like global satellite navigation system (GNSS) and satellite communication benefit from the use of CP antennas. One key parameter is AR bandwidth. The working principle for enhancing AR bandwidth with feeding networks is discussed in this section, such as feeding scheme with resonator types in Section 5.4.1, feeding scheme with unequal resonator numbers in Section 5.4.2, and feeding scheme with different coupling structures in Section 5.4.3.

5.4.1 Feeding with Different Resonator Types

In this section, a wideband CP antenna is designed using a codesign approach for the patch radiator and feeding network. It employs one $\lambda/4$ resonator and one $\lambda/2$ resonator to simultaneously excite two orthogonal modes of a patch antenna [19]. As two impedance-matching elements, they provide a nearly equal magnitude of two orthogonal modes in a wide-frequency band, and result in a desired 90°-phase difference. A T-junction without the power-consuming resistor is connected with these two resonators, and wideband AR response with two minima is obtained.

Figure 5.18 shows configuration and fabrication photographs of the proposed CP patch antenna, which is in a multilayer structure formed by two substrates and three metallic layers. A square patch radiator is printed on layer 1, and a microstrip feeding network is printed on layer 2. The feeding network consists of one $\lambda/4$ resonator, one $\lambda/2$ resonator, and an equal-magnitude and in-phase T-junction. These two resonators are placed along the *x*- and *y*-direction central lines beneath the patch to excite the TM_{10} and TM_{01} modes, respectively. The metallic via-hole provides a short-circuited ground to the $\lambda/4$ resonator and functions as a coupling element connected to one output line of T-junction. Another output line of T-junction is coupled to the $\lambda/2$ resonator through edge coupling. The equivalent circuit of the proposed wideband CP antenna is also provided in Figure 5.18.

A proposed CP antenna is designed and analyzed with the equivalent-circuit model in Figure 5.18. Both two orthogonal modes of the patch radiator are modeled as parallel GLC (G_A, L_A, and C_A) resonant circuits, as presented in many designs of filtering antennas [20–22]. In this case, the radiation loss of the patch antenna is embodied by the radiation conductance G_A, and the magnitude and phase of radiated far field can be directly mapped to the output on G_A in the circuit model. Proximity couplings between the line resonators and patch radiator are equivalently expressed as J inverters with –90° phase shift. Two output lines of the T-junction are coupled to the $\lambda/4$ and $\lambda/2$ resonators through a K inverter and a J inverter, respectively. The $\lambda/4$ resonator and the TM_{10} mode of the patch radiator form a second-order filter network, while the $\lambda/2$ resonator and the TM_{01} mode of the patch radiator make up another second-order filter network. These two filter networks will be designed later on according to the Chebyshev low-pass prototype.

To obtain all the element values in equivalent-circuit model, the dimensions of the patch radiator are determined. Substrate 1 is Rogers 5870 with thickness of 1.575 mm and relative permittivity of 2.33; substrate 2 is Rogers 4003C with thickness of 0.813 mm and relative permittivity of 3.55. The overall size of the patch radiator is 16 mm × 16 mm at the resultant resonant frequency of 5.15 GHz. The values of G_A, L_A, and C_A can be calculated in the closed-form manner based on the cavity-model theorem [23]; they are G_A = 3.87 mS, L_A = 0.597 nH, and C_A = 1.6 pF, respectively.

Referring to design tables in [24], the normalized element values of the Chebyshev low-pass filter prototype with a ripple level of 0.05 dB are $g_0 = 1$,

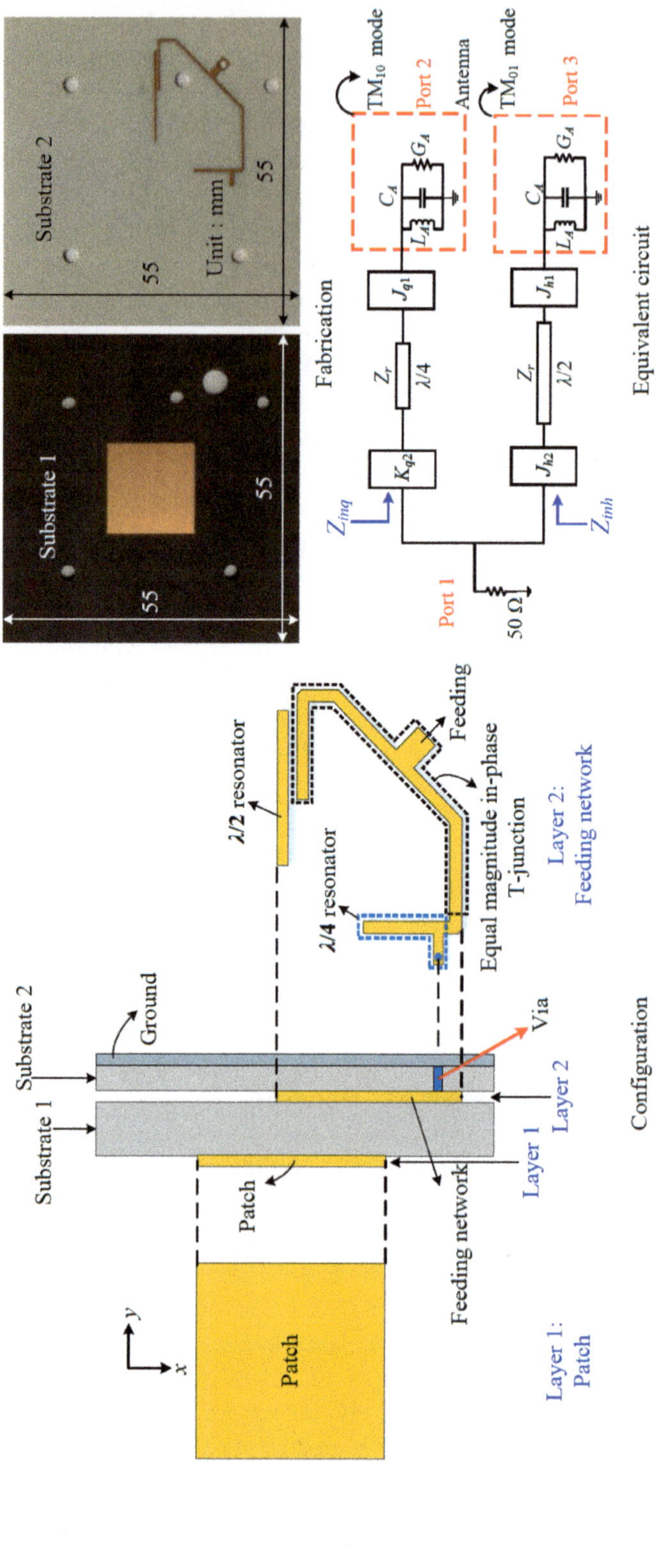

Figure 5.18 Configuration, fabrication, and equivalent-circuit model of the proposed wideband CP patch antenna.

$g_1 = 0.6923$, $g_2 = 0.5585$, and $g_3 = 1.2396$. The values of J and K inverters of the equivalent-circuit model can be calculated as follows [24]

$$Q_A = \frac{\omega_0 C_A}{G_A} \tag{5.11}$$

$$FBW = \frac{g_2 g_3}{Q_A} \tag{5.12}$$

$$J_{q1} = FBW\sqrt{\frac{\pi\omega_0 C_A}{4g_1 g_2 Z_r}} \tag{5.13}$$

$$K_{q2} = \sqrt{\frac{FBW\pi Z_r Z_0}{4g_0 g_1}} \tag{5.14}$$

$$J_{h1} = FBW\sqrt{\frac{\pi\omega_0 C_A}{2g_1 g_2 Z_r}} \tag{5.15}$$

$$J_{h2} = \sqrt{\frac{\pi FBW}{2g_0 g_1 Z_0 Z_r}} \tag{5.16}$$

where Z_r and Z_0 are the characteristic impedance of the resonators and the reference impedance of the networks, and their values are 96Ω and 100Ω, respectively.

With the use of (5.11) to (5.16), the resultant values of the inverters can be derived as $J_{q1} = 1.71$ mS, $K_{q2} = 23.74\Omega$, $J_{h1} = 2.42$ mS, and $J_{h2} = 3.50$ mS. The parallel GLC circuits of two orthogonal modes can be matched to Z_0 (100Ω) after connecting the $\lambda/4$ and $\lambda/2$ resonators.

Input impedances (Z_{inh} and Z_{inq}) of the two filter networks are calculated and plotted in Figure 5.19. Due to the distinctive characteristics of $\lambda/4$ and $\lambda/2$ resonators, the real parts of Z_{inh} and Z_{inq} vary in different oscillating ways, showing ripples around reference impedance $Z_0 = 100\Omega$. Meanwhile, their

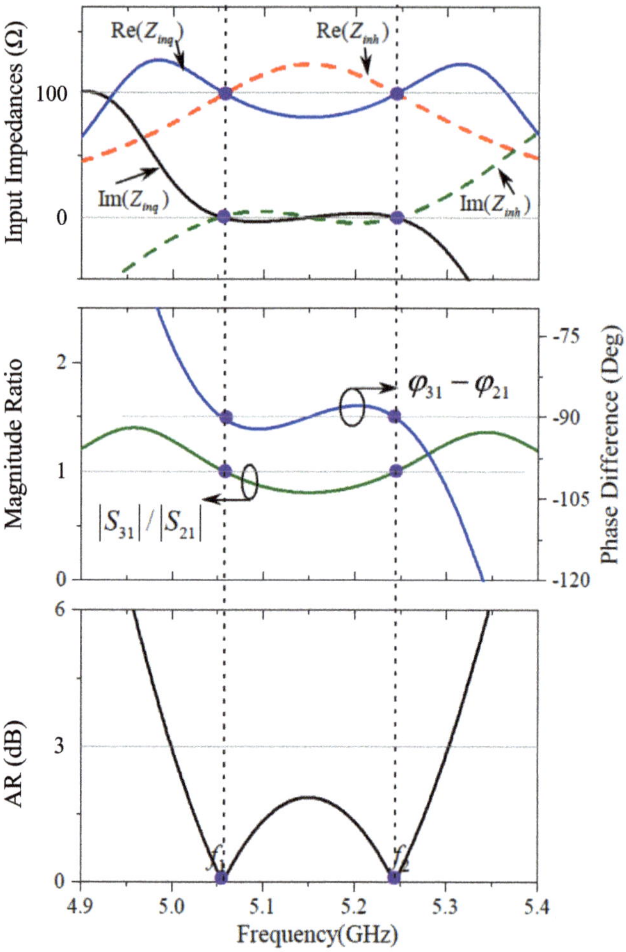

Figure 5.19 Theoretical input impedances, magnitude ratio, phase difference, and predicted AR response based on the equivalent-circuit model in Figure 5.18.

imaginary parts keep negligible variation around zero near the center frequency. At two symmetric frequency points of f_1 and f_2, both Re(Z_{inh}) and Re(Z_{inq}) have the same value of 100Ω, while Im(Z_{inh}) and Im(Z_{inq}) are exactly zero. The input impedance of two branches becomes very close to each other within the frequency band of interest, which will further lead to the desired magnitude of the two modes.

The magnitude ratio ($|S_{31}|/|S_{21}|$) and phase difference ($\phi_{31} - \phi_{21}$) are calculated, and the results are plotted in Figure 5.19. Due to the closely equalized input impedance Z_{inh} and Z_{inq}, the magnitude ratio is quite flat, and its value slightly varies around 1. Due to the difference in $\lambda/4$ and $\lambda/2$ resonators, the phase difference near the center frequency is kept stable around –90°. At two frequency points of f_1 and f_2, the two orthogonal modes will be excited with exactly the same magnitude and quadrature phase, and the input impedance at the T-junction is thus satisfactorily matched to 50Ω.

Based on the antenna theory in [25], it can be deduced that the AR, $|S_{31}|/|S_{21}|$, and $\phi_{31} - \phi_{21}$, should be strictly satisfied with the following relationship:

$$\gamma = \tan^{-1}\left(|S_{31}| / |S_{21}|\right) \tag{5.17}$$

$$\delta = \varphi_{31} - \varphi_{21} \tag{5.18}$$

$$\varepsilon = 0.5\sin^{-1}\left(\sin(2\gamma)\sin\delta\right) \tag{5.19}$$

$$AR = 1/\tan\varepsilon \tag{5.20}$$

Using (5.17) to (5.20), the theoretical AR responses can be calculated and the relevant results shown in Figure 5.19. Two minima with values of 0 dB in the AR response come out at frequency points of f_1 and f_2 as a result of equal magnitude and quadrature phase of the two modes.

Figure 5.20 shows the two sets of simulation and measured reflection coefficients. The measured frequency range of the proposed antenna with $|S_{11}| <$ -10 dB is 4.92 to 5.4 GHz. The realized gain is flat in the operating band and sharply rolls down out of band, showing wideband and filtering characteristics. Simulated and measured efficiencies of the two antennas are depicted in Figure 5.20, and they show good agreement. Compared to a traditional antenna, the proposed CP patch antenna has wider bandwidth and flatter response in band. Simulated and measured AR responses are plotted in Figure 5.20. For the proposed CP antenna, two minima come up with a ripple, generating much wider AR bandwidth. The simulated and measured fractional bandwidths of the proposed antenna with AR < 3 dB are 5.4% and 6.2%, respectively.

In summary, one $\lambda/4$ resonator and one $\lambda/2$ resonator are employed to simultaneously excite two orthogonal modes of the patch antenna. The amplitude and phase conditions required for wideband CP radiation are simultaneously satisfied in a codesign by carrying out the two feeding networks as second-order filter, increasing the AR bandwidth of the antenna as desired. The proposed multimode resonance-inspired antenna achieved the wide AR bandwidth, compact size, stable beam, and filtering gain response.

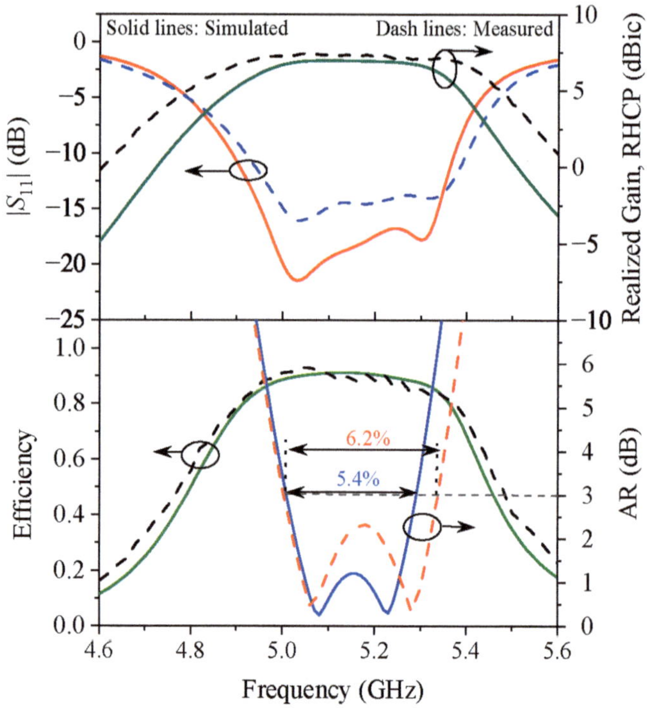

Figure 5.20 Simulated and measured reflection coefficients, realized gain, efficiencies, and AR responses of the proposed CP patch antennas in Figure 5.18.

5.4.2 Feeding with Different Resonator Numbers

In Section 5.4.1, a 90°-phase difference is achieved by using different type resonators. In this section, the phase condition is satisfied by employing a different resonator number in the feeding networks [26]. Figure 5.21 shows configuration and fabrication photographs of the proposed CP MPA, where a square patch radiator and the feeding network lay on the upper surface of substrate 1 and substrate 2, respectively. The feeding network includes coupled $\lambda/2$ resonators and an equal-magnitude and in-phase Wilkinson power divider; and it feeds the patch through proximity coupling. Through the power divider, the incident wave is divided in two parts with equal magnitude and phase, which flow along path 1 and path 2 to excite two orthogonal modes of the patch. Path 1 and path 2 consist of one and two $\lambda/2$ resonators, respectively. The power divider is connected to the resonators by tap coupling. The resonators coupled to the patch are aligned along the central lines of the patch in the *x*- and *y*-directions. Along the two paths, path 2 consists of one more *J* inverter and one more $\lambda/2$ resonator. It would produce –90°-phase shift, which is required for CP radiation.

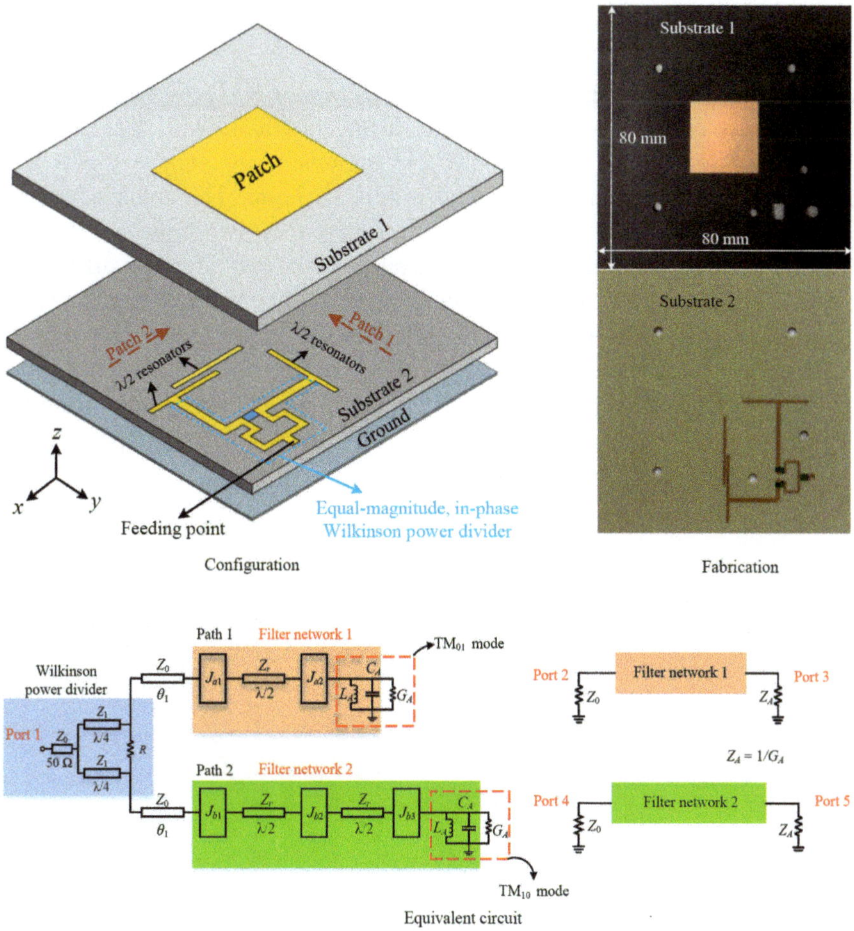

Figure 5.21 Configuration, fabrication, and equivalent-circuit model of the proposed wideband CP MPA.

The equivalent-circuit network of the proposed CP antenna is shown in Figure 5.21. A parallel GLC resonator circuit (G_A, L_A, and C_A) is used to model each of the two orthogonal modes. Couplings between resonators or between resonator and patch are represented as J inverters with –90°-phase shift. Two bandpass filter networks are formed, and the radiating patch resonator acts as the last-stage resonator and output port of these two filter networks. The patch antenna radiation loss is embodied by the radiation conductance G_A, and the magnitude and phase of far field can be directly mapped to the output on G_A in the circuit model. The magnitudes and phases of excitation of TM_{01} and TM_{10} modes will be synthesized at the same time.

Since the two output ports of the Wilkinson power divider are well isolated, the whole design can be further simplified into two filter networks. The impedances at two ports of each filter network are Z_0 and Z_A, and Z_A stands for the radiation resistance of a patch antenna.

Substrate 1 is Rogers 5870 with a thickness of 1.575 mm and a relative permittivity of 2.33, while substrate 2 is Rogers 4003C with a thickness of 0.508 mm and a permittivity of 3.55. The resonant frequency is 4 GHz, and the resultant patch dimension is 21.6 mm × 21.6 mm. The unloaded quality factor (Q_0) of the patch under the specification of these two substrates is 22.25.

The reflection and transmission response of the two filter networks in Figure 5.21 can be derived and calculated by applying the theory of bandpass filter in [24]. The detailed derivation and analysis are provided in our published work in [26], and the theoretical results are shown in Figure 5.22. The magnitude ratio and phase difference of S_{32} and S_{54} vary in an oscillating way with a few ripples between f_1 and f_2. As a result, the AR turns out to have ripple response and shows three minima at f_0, f_1, and f_2.

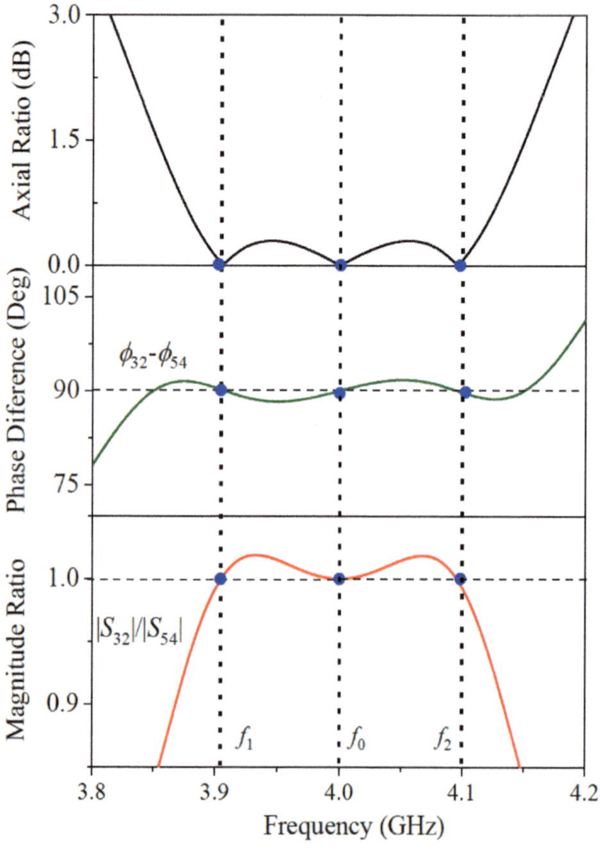

Figure 5.22 Theoretical magnitude ratio, phases difference, and predicted AR responses for f_1 = 3.905 GHz.

With the above arrangements, the required dimensions of the patch at a resonant frequency of 4 GHz should be 21.6 mm × 21.6 mm, and the unloaded quality factor is equal to 22.25. Filter network 1 is designed to exhibit the Chebyshev response with a ripple level of 0.3 dB, and the resulting external quality factor and coupling coefficient are $Q_{ea} = Q_0 = 22.25$ and $k_{12a} = 0.05854$, respectively. When frequency f_1 is specified as 3.905 GHz, the parameters of filter network 2 can be calculated as $Q_{eb} = 5.46$, $k_{12b} = 0.1026$, and $k_{23b} = 0.0662$. The definition and meaning of these quality factors and coupling coefficients can be found in [26].

As a result of the prescribed response in Figure 5.23, the simulated AR of the antenna in Figure 5.21 agrees with the theoretical one. Low-ripple response and three minima of AR within a bandwidth can be seen, and the AR ascends sharply outside the frequency range of interest. The FBW of the measured AR lower than 3 dB is about 8.8%, which is much wider than that of a traditional single-fed CP patch antenna.

Simulated and measured gains and reflection coefficients are plotted in Figure 5.23. There is a flat gain response over a frequency band around the center frequency, and the realized gain rolls off at two sides of the frequency

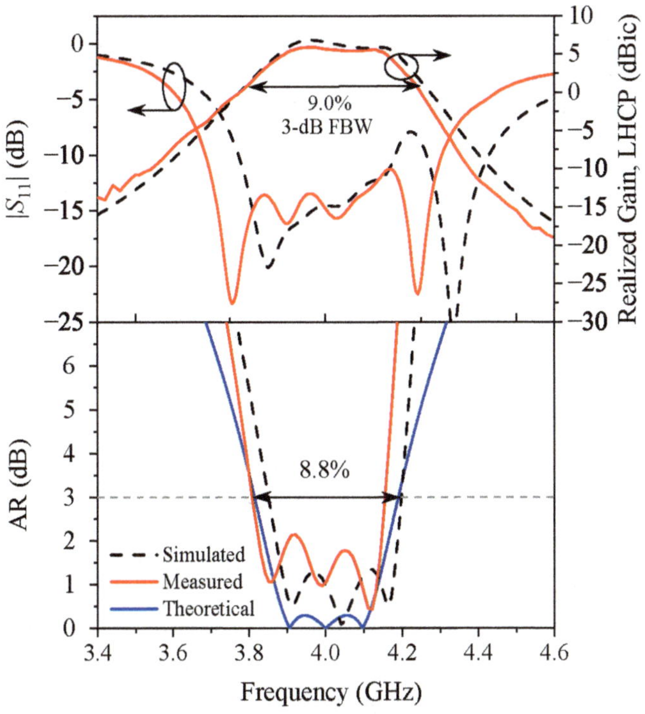

Figure 5.23 Simulated, measured, and theoretical AR responses, realized gains, and reflection coefficients of the designed antenna in Figure 5.21.

band, showing good filtering characteristics. The measured frequency range of 3-dB-gain variation is from 3.84 to 4.2 GHz with FBW of 9%.

By using one resonator and two $\lambda/2$ resonators at the two feeding paths to construct two dissimilar bandpass filter networks with different orders, the AR bandwidth of the antenna could be further improved with three minimum poles. The proposed multimode resonance-inspired antenna has achieved the wide AR bandwidth, compact size, stable beam, and filtering gain response.

5.4.3 Feeding with Different Coupling Structure

In Sections 5.4.1 and 5.4.2, the CP antennas consist of multilayer schemes, which will increase their structural complexity and fabrication cost. To address this issue, the tapped line and parallel coupled line are both employed and realized on a single-layer substrate. In this design, the 90°-phase shift is obtained through the difference between the phase response of tapped line and parallel coupled line [27].

The geometry of the proposed antenna and a photograph of the fabricated prototype with detailed dimensional parameters are shown in Figure 5.24, where two $\lambda/2$ resonators are employed to excite the patch radiator to achieve wideband magnitude response of two orthogonal polarizations. Two resonators are coupled to the feeding line, and these two coupling structures can provide the required 90°-phase difference for CP radiation. An equal-magnitude T-junction is formed to link two paths with the input port.

The equivalent circuit of the whole antenna is shown in Figure 5.24, where the patch radiator is modeled as a parallel GLC resonator consisting of conductance, inductance, and capacitance. The coupling between the $\lambda/2$ resonator and patch radiator is represented by a J inverter. The feeding circuits and radiator in two paths can be jointly designed as second-order filter networks. Though coupling schemes of feeding are different, the two filter networks can be carried out with the same Chebyshev low-pass filter prototype to achieve the same ripple level and bandwidth. The equivalent-circuit model for exciting TM_{10} mode contains an additional J inverter, J_{01}, so it will produce the 90°-phase shift for CP radiation.

By using a similar method in Sections 5.4.1 and 5.4.2, the real parts of two input impedances are calculated to be around 100Ω at the operating band, as shown in Figure 5.25, while the imaginary parts are around 0Ω. At two frequencies, f_1 and f_2, input impedances are exactly 100Ω. The magnitude ratio and phase difference are around 1 and 90°, respectively, with small ripples within the operating band, which satisfy the required magnitude and phase conditions for CP radiation. A wide 3-dB AR bandwidth is achieved in Figure 5.25. At f_1 and f_2, two minima of 0 dB in the AR response exist, where the magnitude ratio and phase difference are 1 and 90°, respectively.

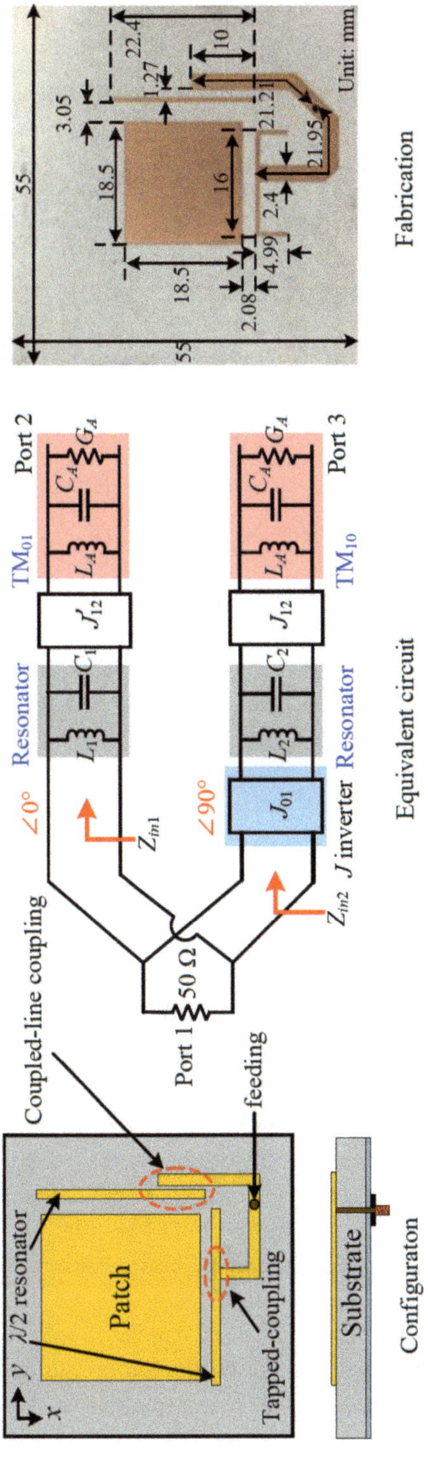

Figure 5.24 Configuration, fabrication, and equivalent-circuit model of the proposed CP patch antenna.

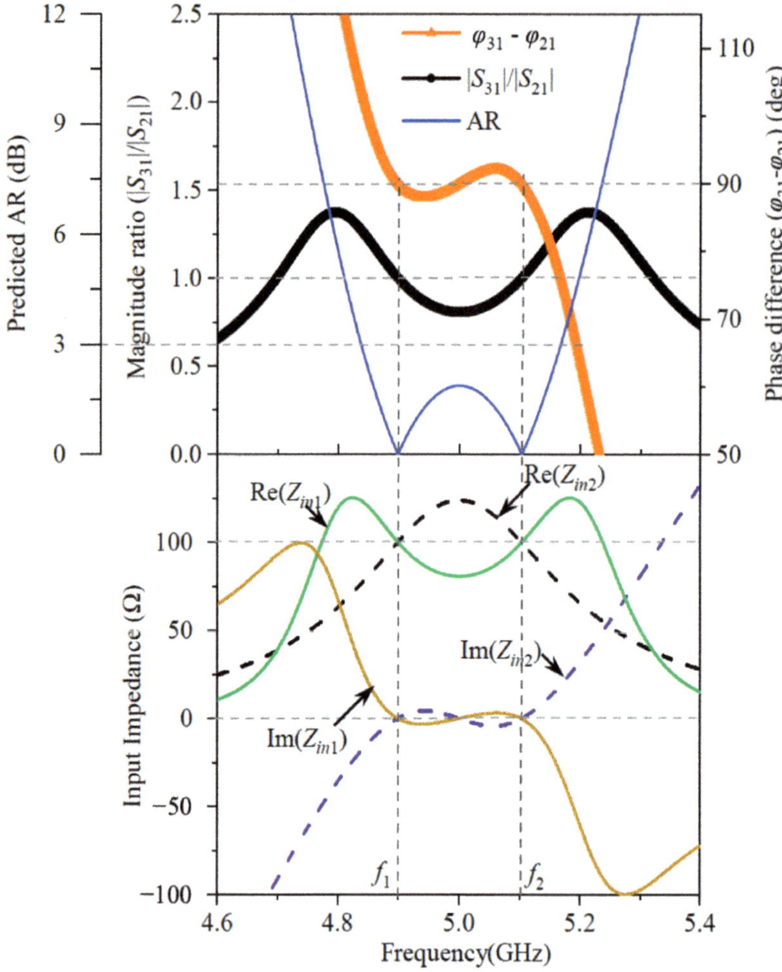

Figure 5.25 Input impedances from calculation and theoretical magnitude ratio, phase difference, and AR responses based on the equivalent-circuit model in Figure 5.24.

Simulated and measured $|S_{11}|$ with realized gain are shown in Figure 5.26. There is a flat range with a wide bandwidth in the realized gain of the proposed CP patch antenna. Figure 5.26 also shows the simulated and measured AR responses of the proposed CP patch antennas at broadside. Two minima merge to form a wide AR bandwidth. The simulated and measured 3-dB AR fractional bandwidths are 7.0% and 7.07%, respectively.

In summary, by using tapped line and parallel coupled line, the AR-bandwidth of the MPA has been enhanced with two minimum poles. Most important, wideband CP radiation is realized in a single layer. The proposed multimode resonance-inspired antenna has the advantages of wide AR bandwidth,

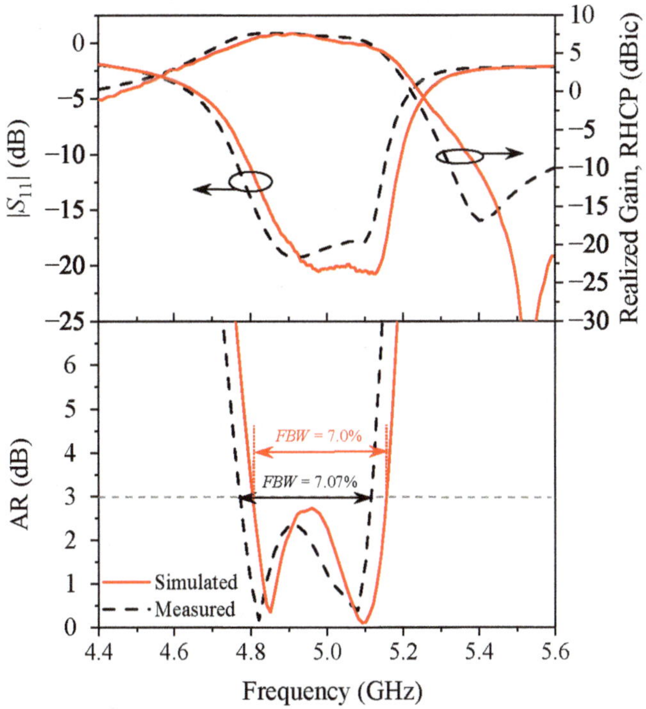

Figure 5.26 Simulated and measured $|S_{11}|$, realized gains, and AR response.

compact size, single-layer implementation, stable beam, and filtering gain response.

In addition to the above methods of increasing AR bandwidth, other feature designs are shown in Figure 5.27. In [28], four parasitic patches are introduced and coupled to the four driven patches, which generates two CP modes and achieves a 3-dB AR bandwidth of 12.9%. In [29], two orthogonal modes of the patch are excited via two coupling paths in a single resonator. By exploiting the different orders of the two coupling paths, a coherent 90°-phase difference is produced and a 3-dB AR bandwidth of 21.2% at X-band. In [30], by loading an asymmetric U-slot in the patch, three orthogonal linearly polarized modes are generated, with the center mode shared by two cross-polarized modes of the same polarization, resulting in a 3-dB AR bandwidth of 21%.

5.5 Conclusions

Antenna radiation gain can be enhanced in several ways. For example, broadside gain can be enhanced by using an array technique, and CP performance can be improved by using a traditional feeding network. But compared to these

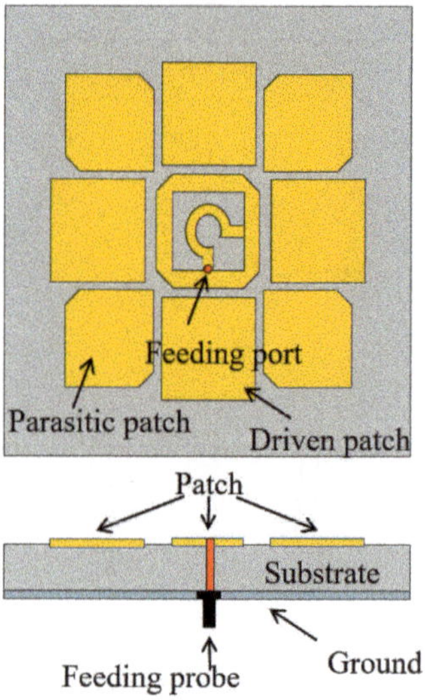

[28]

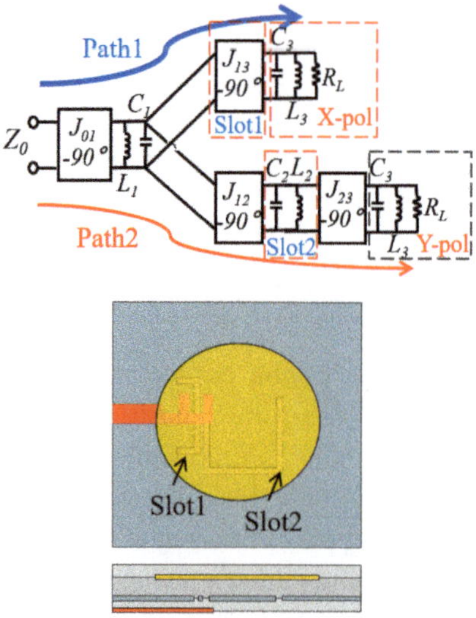

[29]

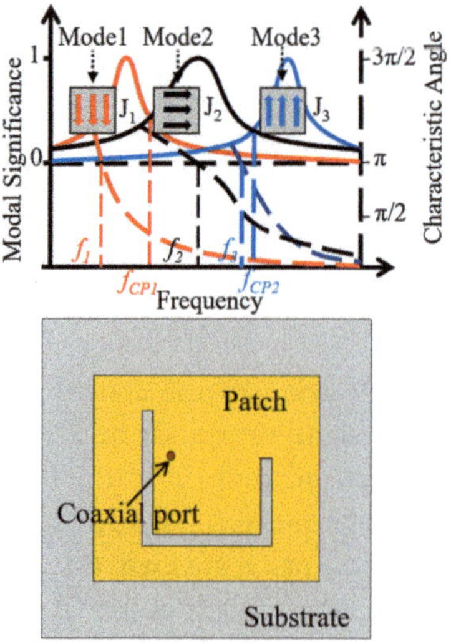

[30]

Figure 5.27 Illustration of other AR bandwidth enhancement methods.

abovementioned methods, the multimode resonance-inspired antennas with gain improvement in this section have the advantages of compact size, stable gain, consistent radiation beam, simple structure, and so on. These proposed antennas have the potential applications in the point-to-point communication, GNSS, and filtering systems.

References

[1] Zhang, X., and L. Zhu, "Gain-Enhanced Patch Antennas with Loading of Shorting Pins," *IEEE Transactions on Antennas and Propagation,* Vol. 64, No. 8, August 2016, pp. 3310–3318.

[2] Garg, R., et al., *Microstrip Antenna Design Handbook,* Norwood, MA: Artech House, 2001.

[3] Balanis, C. A, *Antenna Theory: Analysis and Design,* New York: Wiley, 2005.

[4] Zhang, X., and L. Zhu, "Gain-Enhanced Patch Antenna Without Enlarged Size Via Loading of Slot and Shorting Pins," *IEEE Transactions on Antennas and Propagation,* Vol. 65, No. 11, November 2017, pp. 5702–5709.

[5] Zhang, X., and L. Zhu, "Dual-Band High-Gain Differentially Fed Circular Patch Antenna Working in TM_{11} and TM_{12} Modes," *IEEE Transactions on Antennas and Propagation,* Vol. 66, No. 6, June 2018, pp. 3160–3165.

[6] Juyal, P., and L. Shafai, "Sidelobe Reduction of TM_{12} Mode of Circular Patch Via Nonresonant Narrow Slot," *IEEE Antennas and Wireless Propagation Letters,* Vol. 64, No. 8, August 2016, pp. 3361–3369.

[7] Liu, J., et al., "Broadband Monopolar Microstrip Patch Antenna with Shorting Vias and Coupled Ring," *IEEE Antennas and Wireless Propagation Letters,* Vol. 13, 2014, pp. 39–42.

[8] Squadrito, P., S. Zhang, and G. F. Pedersen, "Wideband or Dual-band Low-Profile Circular Patch Antenna with High-Gain and Sidelobe Suppression," *IEEE Transactions on Antennas and Propagation,* Vol. 66, No. 6, June 2018, pp. 3166–3171.

[9] Luo, Y., Z. -N. Chen, and K. Ma, "A Single-Layer Dual-Polarized Differentially Fed Patch Antenna with Enhanced Gain and Bandwidth Operating at Dual Compressed High-Order Modes Using Characteristic Mode Analysis," *IEEE Transactions on Antennas and Propagation,* Vol. 68, No. 5, May 2020, pp. 4082–4087.

[10] Juyal, P., and L. Shafai, "A Novel High-Gain Printed Antenna Configuration Based on TM_{12} Mode of Circular Disc," *IEEE Transactions on Antennas and Propagation,* Vol. 64, No. 2, February 2016, pp. 790–796.

[11] Zhang, X., et al., "Resonator-Fed Wideband and High-Gain Patch Antenna with Enhanced Selectivity and Reduced Cross-Polarization," *IEEE Access,* Vol. 7, 2019, pp. 49918–49927.

[12] Zhang, X., L. Zhu, and Q. -S. Wu, "Sidelobe-Reduced and Gain-Enhanced Square Patch Antennas with Adjustable Beamwidth Under TM_{03} Mode Operation," *IEEE Transactions on Antennas and Propagation,* Vol. 66, No. 4, April 2018, pp. 1704–1713.

[13] Hong, J. -S., and M. J. Lancaster, *Microstrip Filters for RF/Microwave Applications,* New York: Wiley, 2001.

[14] Zhang, X., et al., "Wideband Differentially Fed Patch Antennas Under Dual High-Order Modes for Stable High Gain," *IEEE Transactions on Antennas and Propagation,* Vol. 69, No. 1, January 2021, pp. 508–513.

[15] Won, K. -L., and T. -W. Chiou, "Broad-band Single-Patch Circularly Polarized Microstrip Antenna with Dual Capacitively Coupled Feeds," *IEEE Transactions on Antennas and Propagation,* Vol. 49, No. 1, January 2001, pp. 41–44.

[16] Wang, Z., J. Liu, and Y. Long, "A Simple Wide-Bandwidth and High-Gain Microstrip Patch Antenna with Both Sides Shorted," *IEEE Antennas and Wireless Propagation Letters,* Vol. 18, No. 6, June 2019, pp. 1144–1148.

[17] Yang, W., et al., "Novel Compact High-Gain Differential-Fed Dual-Polarized Filtering Patch Antenna," *IEEE Antennas and Wireless Propagation Letters,* Vol. 67, No. 12, December 2019, pp. 7261–7271.

[18] Liu, G., et al., "Bandwidth and Gain Enhancement of a Single-Layer Filtering Patch Antenna Using Reshaped TM_{12} Mode," *IEEE Antennas and Wireless Propagation Letters,* Vol. 23, No. 1, January 2024, pp. 314–318.

[19] Wu, Q. -S., X. Zhang, and L. Zhu, "A Wideband Circularly Polarized Patch Antenna with Enhanced Axial Ratio Bandwidth Via Codesign of Feeding Network," *IEEE Transactions on Antennas and Propagation,* Vol. 66, No. 10, October 2018, pp. 4996–5003.

[20] Pues, H. -F., and A. R. van de Capelle, "An Impedance-Matching Technique for Increasing the Bandwidth of Microstrip Antennas," *IEEE Transactions on Antennas and Propagation,* Vol. 37, No. 11, November 1989, pp. 1345–1354.

[21] Le Nadan, T., et al., "Integration of an Antenna/Filter Device, Using a Multilayer, Multitechnology Process," *28th European Microwave Conference,* Amsterdam, Netherlands, October 1998, pp. 672–677.

[22] Jiang, Z. -H., and D. H. Werner, "A Compact, Wideband Circularly Polarized Codesigned Filtering Antenna and Its Application for Wearable Devices with Low SAR," *IEEE Transactions on Antennas and Propagation,* Vol. 63, No. 9, September 2015, pp. 3808–3818.

[23] Garg, R., et al., *Microstrip Antenna Design Handbook,* Norwood, MA: Artech House, 2001.

[24] Matthaei, G. L., L. Young, and E. M. T. Jones, *Microwave Filters, Impedance-Matching Networks and Coupling Structures,* Norwood, MA: Artech House, 1980.

[25] Kraus, J. D., and R. J. Marhefka, *Antennas for All Applications,* Third Edition, New York: McGraw-Hill, 2001.

[26] Wu, Q. -S., X. Zhang, and L. Zhu, "Codesign of a Wideband Circularly Polarized Filtering Patch Antenna with Three Minima in Axial Ratio Response," *IEEE Transactions on Antennas and Propagation,* Vol. 66, No. 10, October 2018, pp. 5022–5030.

[27] Wu, Q. -S., X. Zhang, and L. Zhu, "A Feeding Technique for Wideband CP Patch Antenna Based on 90° Phase Difference Between Tapped Line and Parallel Coupled Line," *IEEE Antennas and Wireless Propagation Letters,* Vol. 18, No. 7, July 2019, pp. 1468–1471.

[28] Ding, K., et al., "Compact Broadband Circularly Polarized Antenna with Parasitic Patches," *IEEE Transactions on Antennas and Propagation,* Vol. 65, No. 9, September 2017, pp. 4854–4857.

[29] Mao, C. -X., et al., "Dual-Band Circularly Polarized Shared-Aperture Array for C-/X-band Satellite Communications," *IEEE Transactions on Antennas and Propagation,* Vol. 65, No. 10, October 2017, pp. 5171–5178.

[30] Zeng, J., et al., "Single-Fed Triple-Mode Wideband Circularly Polarized Microstrip Antennas Using Characteristic Mode Analysis," *IEEE Transactions on Antennas and Propagation,* Vol. 70, No. 2, February 2022, pp. 846–855.

6

Multimode Resonance-Inspired Planar Antennas with Pattern-Bandwidth Improvement

6.1 Introduction

The main focus of this chapter is to show methods for improving the performance of planar antennas through pattern bandwidth. HPBW-enhancement techniques will be introduced in Section 6.2; sidelobe level reduction techniques will be the discussion in Section 6.3; cross-polarization reduction will be illustrated in Section 6.4; and the beam scanning technique will be illustrated in Section 6.5.

Previous works have been proposed and developed, such as using parasitic vertical element [1, 2], high impedance surface [3], and multimode concept [4]. In the following, the working principle on pattern-bandwidth manipulation will be extensively discussed.

6.2 Wide HPBW Bandwidth Technique

In wireless communication systems, antennas with wide half-power beamwidth (HPBW) or axial-ratio (AR) beamwidth are often used because of their wide signal coverage. Unfortunately, traditional planar antennas suffer from a narrow beam. For example, the HPBW of a MPA is restricted to around 90° or narrower in both E- and H-planes. Several methods have been proposed and developed for the antennas during the past two decades, such as using a single equivalent magnetic current, shown in Section 6.2.1, using a dielectric constant technique, shown in Section 6.2.2, and using a traveling-wave-fed slot spiral, shown in Section 6.2.3.

6.2.1 Single Equivalent Magnetic Current

One effective method to address narrow HPBW is to use a single equivalent magnetic current instead of two parallel equivalent magnetic currents. In the following, we will illustrate this technique in detail.

Figure 6.1 shows the schematic of a proposed MPA composed of a radiating patch, shorting vias, and ground plane. Its dielectric substrate is F4B with $\varepsilon_r = 2.2$ and $H = 3$ mm. The shorting vias are loaded around the edges of the radiator to increase HPBW at E-plane. The radiating patch is divided in three parts to realize wide impedance bandwidth under triple-mode resonance. The specific parameters of the antenna are: $L_g = 130$ mm, $L_1 = 19.3$ mm, $L_2 = 18.2$ mm, $L_3 = 21.3$ mm, $L_s = 3.5$ mm, $W_g = 260$ mm, $W_1 = 140$ mm, $W_2 = 69.9$ mm, $D = 5.7$ mm, $D_1 = 1.3$ mm, $R_1 = 1$ mm, $R_2 = 2$ mm, $H = 3$ mm [5].

Evolution of the proposed antenna is introduced in Figure 6.2 to illustrate the design process for HPBW enhancement at E-plane over the wide operating

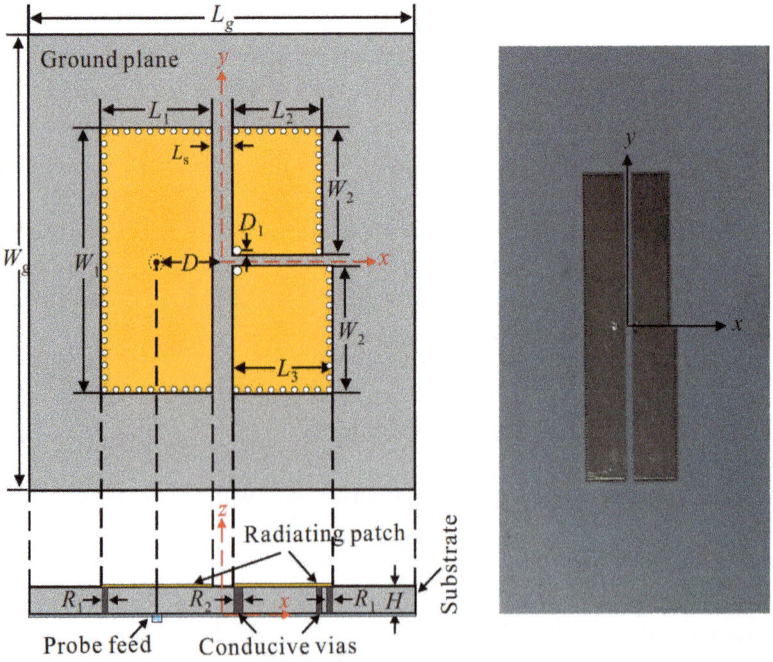

Figure 6.1 Geometry and fabrication of the proposed MPA with wide E-plane beamwidth.

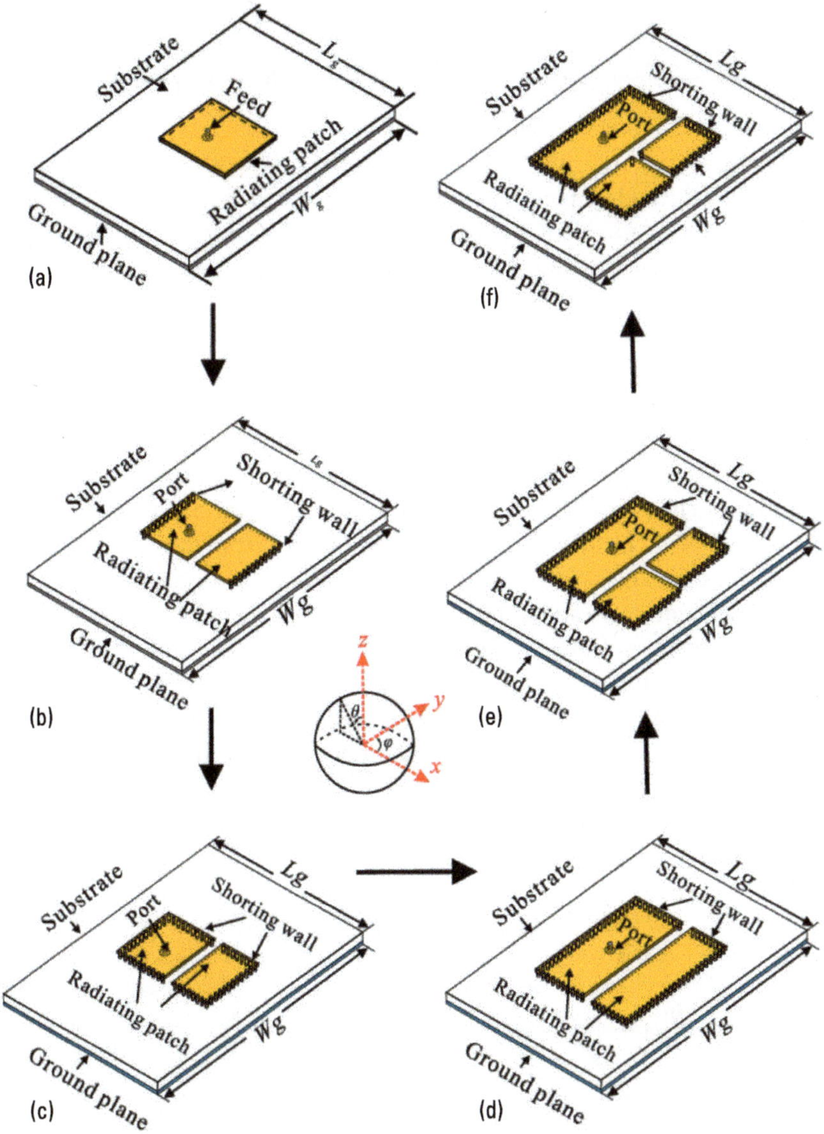

Figure 6.2 Evolution of the patch antenna with E-plane HPBW improvement over the wide operating band: (a) conventional MPA, (b) dual SPAs with loaded pins along x-axial, (c) dual HMCs with loaded pins along x-axial and y-axial, (d) dual HMCs with an enlarged patch width, (e) dual HMCs with an enlarged patch width and a linear slot, and (f) dual HMCs with an enlarged patch width, a linear slot, and conductive vias.

band. First, for the traditional MPA in Figure 6.2(a), it suffers from a narrow HPBW of below 90° under TM_{10} mode at both E-plane and H-plane. Second, the SPA is used in Figure 6.2(b), which shorts two edges of the radiating patch and cuts the slot at the patch center, aiming to remove the array factor and obtain the wide HPBW at *E*-plane pattern. The cross polarization of its *H*-plane radiation pattern is enlarged due to unbalanced radiation edges along *x*-axial. Third, employing a pair of half-mode cavities (HMCs) by shorting the above edges along *x*-axial in Figure 6.2(c), the *E*-plane HPBW of the patch antenna is successfully widened, while keeping the low cross-polarization level. The resonant frequency of the HMCs is pushed up. Fourth, the width of the core radiator along *y*-axial is enlarged as shown in Figure 6.2(d), to tune its resonant frequency down to the desired band. Fifth, in Figure 6.2(e), the patch radiator on the right half is divided in two independent parts to excite the additional resonant mode for further bandwidth enhancement. Unfortunately, its cross polarization at *E*-plane pattern turns up. Sixth, in Figure 6.2(f), two conductive vias are loaded around the corner of the rectangular patches. With these improvements, the proposed antenna could successfully gain several attractive performance characteristics, such as small thickness, wide bandwidth, and desired wide *E*-plane HPBW under the operation of triple modes.

To illustrate the working principle for HPBW enhancement, the far-zone-radiated fields of these above antennas in Figure 6.2(a, c) are discussed, and relevant results plotted in Figure 6.3. According to the cavity-model theory [6], for a traditional MPA, its radiated fields under operation of TM_{10} mode can be calculated and obtained by modeling them as two parallel equivalent magnetic currents. When the substrate thickness H is much smaller than λ_0, the far-zone-normalized electric fields at *E*-plane can be expressed as

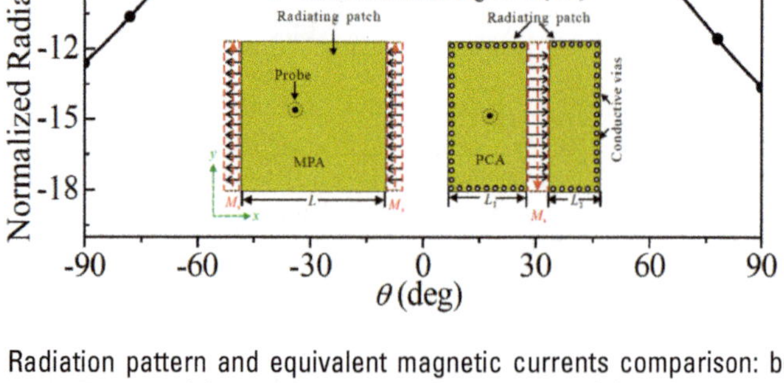

Figure 6.3 Radiation pattern and equivalent magnetic currents comparison: between the traditional MPA in Figure 6.2(a) and improved HMCs in Figure 6.2(c).

$$E_\theta = \cos\left(\frac{k_0 L \sin\theta}{2}\right) \quad (6.1)$$

$$E_\varphi = 0 \quad (6.2)$$

For the improved HMCs in Figure 6.2(c), the far-zone-radiation pattern derived from the primary TM_{10} mode could be also calculated by modeling it as a single equivalent magnetic current due to the shorted two side edges, and its far-zone-normalized electric fields at *E*-plane could be simplified as

$$E_\theta = 1 \quad (6.3)$$

$$E_\varphi = 0 \quad (6.4)$$

As shown in (6.1) and (6.3), the HPBW of the HMCs in Figure 6.2(c) is much wider than that of the traditional MPA in Figure 6.2(a) at E-plane. Meanwhile, Figure 6.3 compares the simulated and calculated radiation patterns between these two antennas. The traditional MPA in Figure 6.2(a) suffers from a narrow *E*-plane HPBW of around 70° and 98° at both finite and infinite ground planes, respectively. Using a pair of HMC schemes, the array factor between two parallel equivalent magnetic currents is successfully removed due to the shorted edges. The HPBW at its *E*-plane pattern could be extended to about 148° and 178° on the finite and infinite ground plane, respectively.

To obtain wide bandwidth under triple resonance, the right core radiator is divided in two parts in Figure 6.2(e), generating three resonant modes over the wideband. However, its *E*-plane cross polarization is increased due to the introduced equivalent magnetic current on the right side. A pair of shorting vias is loaded around the corner of the right patch edges. With these adjustments, wide HPBW, wide bandwidth, and low cross polarization are simultaneously realized.

The proposed antenna in Figure 6.1 was fabricated and measured. As observed in Figure 6.4, the measured and simulated results are depicted and compared: S-parameter, peak realized gain, and E-plane HPBW. The impedance bandwidth of the proposed MPA for $|S_{11}| < -10$ dB could be extended to around 14.3%, which is ranged from 2.54 to 2.93 GHz. It holds a thickness of about $0.027\lambda_0$. The antenna obtains the desired gain and wide *E*-plane HPBW of around 6 dBi and 135° during the operating band, respectively.

In summary, the *E*-plane HPBW can be widened by employing a single equivalent magnetic current from the proposed HMCs instead of the two equivalent magnetic currents from a traditional MPA. The radiating patches

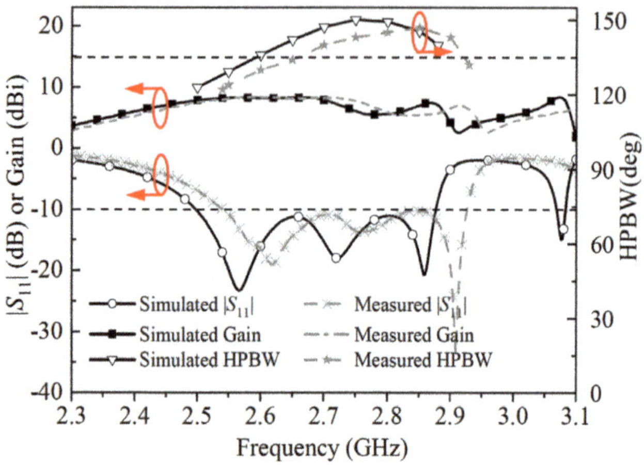

Figure 6.4 Simulated and measured $|S_{11}|$, peak gains, and *E*-plane HPBWs of the fabricated antenna.

are separated as three individual patches to resonate three modes closely to each other for impedance bandwidth enhancement. An array of two shorting vias is inserted at the corner of the radiators, reducing its increased cross polarization at E-plane. The multimode resonant antenna gains low profile, wide bandwidth, and wide *E*-plane HPBW simultaneously.

6.2.2 Dielectric Constant

Apart from the single equivalent magnetic current discussed above, the HPBW of the antenna could be enhanced by controlling the dielectric constant of the substrate. We will illustrate this method in the following. The TM_{10} and TM_{30} modes are simultaneously employed for the antenna with dual-band operation. The TM_{10} mode is excited for E-plane HPBW enhancement in the lower band. The TM_{30} mode is excited for gain enhancement in the upper band.

Figure 6.5 plots the detail configuration of the antenna, which is composed of the patch, center slot, dielectric substrate 1 ($\varepsilon_r = 3.5$ and $H = 4$ mm), dielectric substrate 2 ($\varepsilon_r = 2.2$ and $H = 0.5$ mm), shorting pins, and aperture-coupled feeding. The specific parameters of the antenna are: $L_1 = 50$ mm, $W_1 = 30$ mm, $L_c = 20$ mm, $W_c = 1$ mm, $L_g = 270$ mm, $W_g = 150$ mm, $S_1 = 0.3$ mm, $S_2 = 0.2$ mm, $W_2 = 1.5$ mm, $W_3 = 0.7$ mm, $2R = 0.7$ mm, $L_2 = 9.7$ mm, $L_3 = 11$ mm, $L_4 = 10.5$ mm, $L_5 = 34$ mm, $L_6 = 4.2$ mm, $L_7 = 8.5$ mm, $L_s = 16.9$ mm, $W_s = 1.5$ mm [7].

Initially, based on (6.1) in Section 6.2.1, the E-plane HPBW of a traditional MPA is mainly determined by the spacing between two parallel equivalent magnetic currents. In this view, the key parameter k_0 value is sensitive to

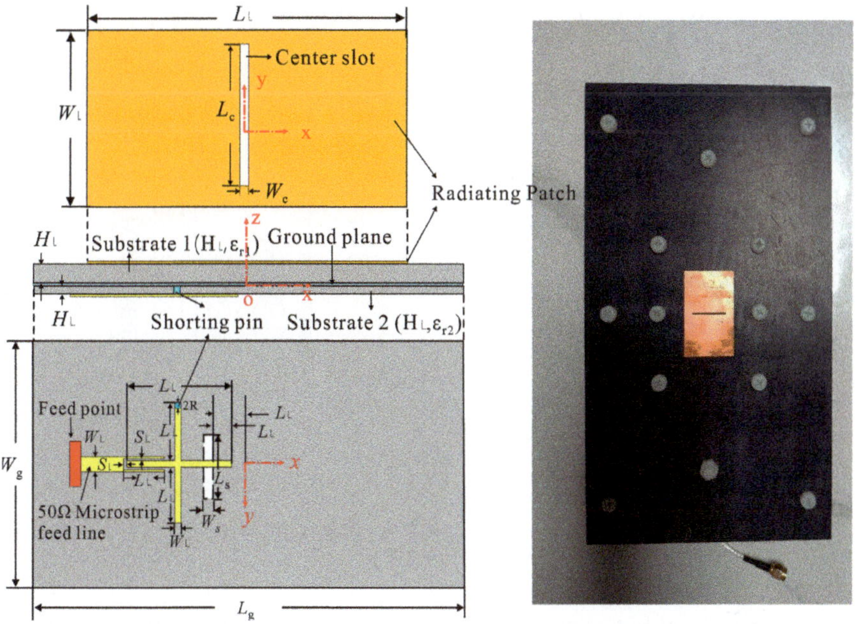

Figure 6.5 Schematic diagram and fabrication of the proposed MPA with wide HPBW at lower band and high gain at upper band via the dielectric constant technique.

the equivalent dielectric constant ε_e. With respect to the traditional MPA, the L/λ_0 under operation of TM_{10} and TM_{30} modes could be established as

$$f_{mn} = \frac{c}{2\sqrt{\varepsilon_e}}\sqrt{\left(\frac{m}{L}\right)^2 + \left(\frac{n}{W}\right)^2} \quad \Rightarrow TM_{10} : \frac{L}{\lambda_{10}} = \frac{1}{2\sqrt{\varepsilon_e}} \tag{6.5}$$

$$f_{mn} = \frac{c}{2\sqrt{\varepsilon_e}}\sqrt{\left(\frac{m}{L}\right)^2 + \left(\frac{n}{W}\right)^2} \quad \Rightarrow TM_{30} : \frac{L}{\lambda_{30}} = \frac{3}{2\sqrt{\varepsilon_e}} \tag{6.6}$$

where L and W are the length and width of the radiating patch by considering the edge effect.

Based on (6.5) and (6.6), the *E*-plane HPBW variation of the antenna at TM_{10} mode and gain response at TM_{30} mode as a function of ε_e is discussed in Figure 6.6. From this figure, by gradually increasing the ε_e, the E-plane HPBW of the antenna under operation of TM_{10} mode could be effectively widened from 120° to about 180°, and the peak gain of the antenna under radiation of TM_{30} mode could be effectively improved from 7.2 to 8.9 dBi. When the ε_e is properly selected as 3.5, the E-plane HPBW of its TM_{10} mode can reach to

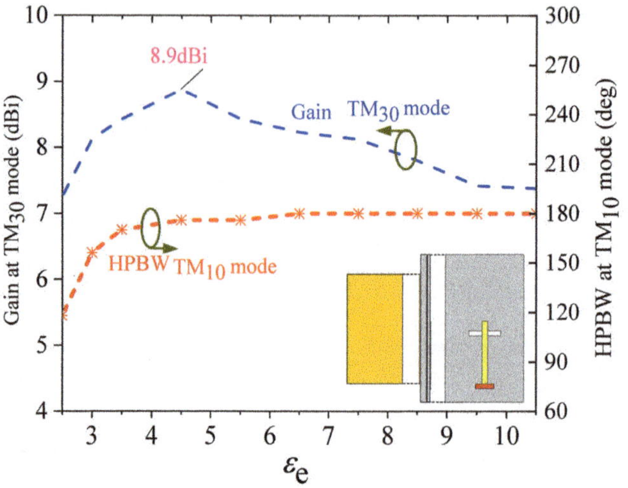

Figure 6.6 *E*-plane HPBW enhancement of TM_{10} mode and gain enhancement of TM_{30} mode under different equivalent dielectric constant ε_e.

around 180°, while keeping an increased gain of around 8.0 dBi at its TM_{30} mode.

The proposed antenna in Figure 6.5 was fabricated and measured. The microstrip feeding line is improved as the coupled line to enhance impedance bandwidth at both lower and upper bands. Meanwhile, the linear slot is cut on the center of the radiating patch for gain enhancement. Figure 6.7 presents the simulated and measured *S*-parameters and gains of the antenna, showing that the designed MPA achieves good impedance matching characteristics in dual working bands of 1.42 to 1.45 GHz, and 3.95 to 4.13 GHz. The antenna keeps stable radiation gain (4.5 dBi at low band, 9.6 dBi at high band) in these dual working band ranges. Wide E-plane HPBW is gained in the lower band.

The high dielectric constant is used to reduce the spacing of TM_{10} mode resonant in parallel current sheets, achieving wide beamwidth in the lower band. Meanwhile, the T-shaped feeding scheme is introduced to realize enhanced bandwidth in dual working bands. The slot is cut on the radiating patch for gain enhancement. The proposed multimode resonance-inspired antenna has the characteristics of low profile, dual band, lower band wide HPBW, higher band high gain, and low cross polarization.

6.2.3 Traveling-Wave-Fed Slot Spiral

The AR beamwidth of CP antennas plays an important role in global navigation satellite systems. In this part, a traveling-wave-fed slot spiral antenna with wide AR beamwidth is proposed and designed. By improving the rotational

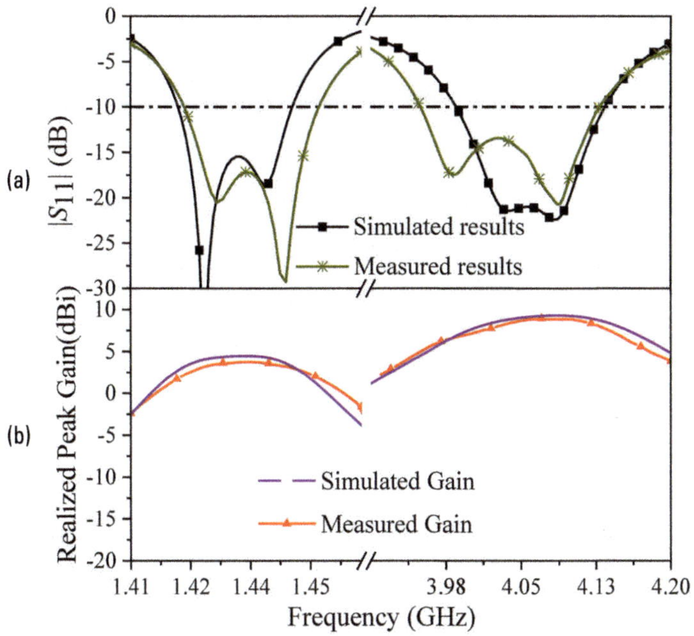

Figure 6.7 Simulated and measured results of the wide HPBW and high-gain MPA based on the dielectric constant technique: (a) S-parameter, and (b) realized peak gain.

symmetry of the antenna with proper number and shape of spiral slot radiators, a better AR uniformity can be obtained in a compact size. The proposed antenna structure in Figure 6.8 consists of 12 identical spiral-radiating slots and two-turn traveling-wave feeding microstrip lines printed on the top and bottom layers of a F4BM substrate ($\varepsilon_r = 2.65$). Four screws fix the substrate to a metallic cavity. Each spiral slot is composed of a radial arm (l_s = 13 mm, w_s = 0.5 mm) and an Archimedean arm (a = 7.5 mm/rad, ϕ_{34} = 150°). The sequential angle between two adjacent radial slots is kept as ϕ_{13} = 30°. The feeding line is divided in two parts: exciting part O_1O_2 (b = 0.5 mm/rad, w_{f2} = 1.1 mm) and matching part OO_1 (w_{f1} = 2 mm). A 50Ω-feeding port and a 150Ω-absorbing resistor are set at O and O_2, respectively. For the antenna configuration shown in Figure 6.8, the "feed port" will be realized by a 50Ω-SMA connector, whose inner pin and outer ground are connected to the feeding line and the metallic plane with etched slots, respectively [8].

Three evolutions are proposed to illustrate the working principle for improving AR performance. The first step is to select the number and shape of slots. Initially, the antenna only consists of two pairs of radial slots in an orthogonal configuration, as shown in case 1 in Figure 6.9. The corresponding AR response at 1.4 GHz in the plane of ϕ = 0° is also illustrated. The AR in this case exhibits low values in the broadside direction. However, the AR in the elevation

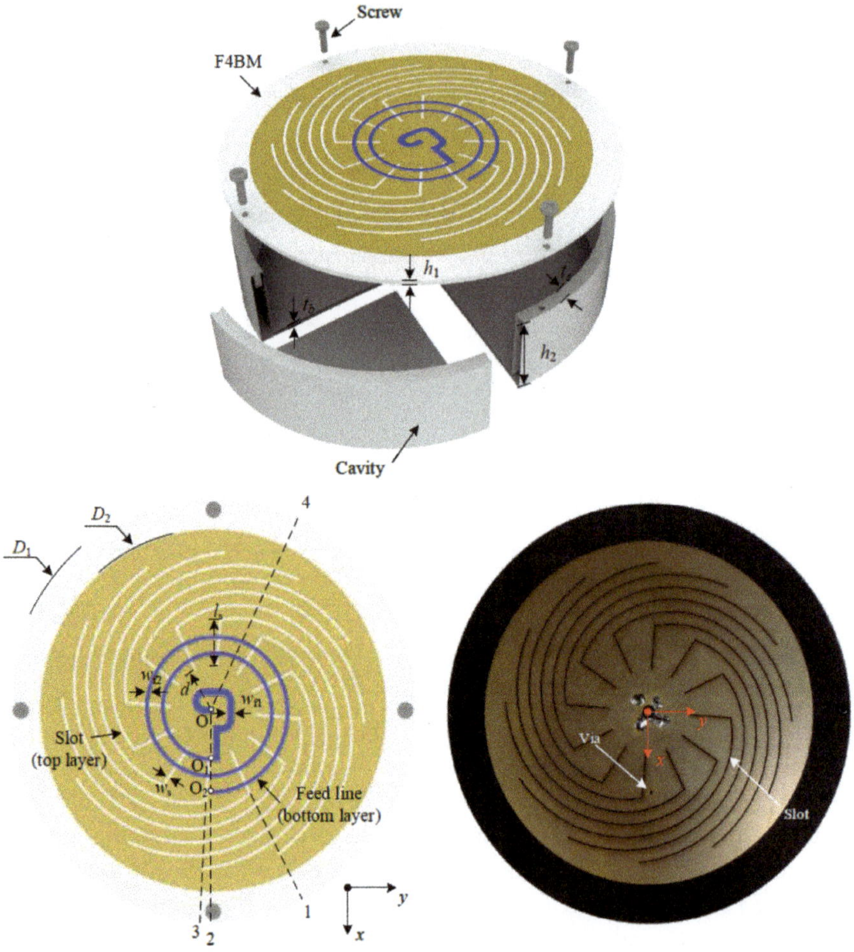

Figure 6.8 Structure and configuration of the proposed antenna.

plane of $\theta = 70°$ undergoes significant variations with pronounced ripples in specific directions, leading to the transformation of the radiated wave into linear polarization. This effect is primarily attributed to the poor rotational symmetry of the fields arising from the substantial angle between adjacent slots. To mitigate the azimuth dependence of the AR response, the slot count is increased to 12 in case 2. This significantly improves AR uniformity, and AR variation is reduced to 2 dB. Finally, by tightly entangling these slots in case 3, the antenna achieves a 55% reduction in area and enhanced AR uniformity of 1 dB.

The next consideration is feeding structure. To activate the 12 spiral slots mentioned earlier, a conventional one-turn traveling-wave feed line, referred to as case A in Figure 6.9, is employed. However, the feeding scheme exhibits significant frequency dependency in both magnitude and phase responses, so

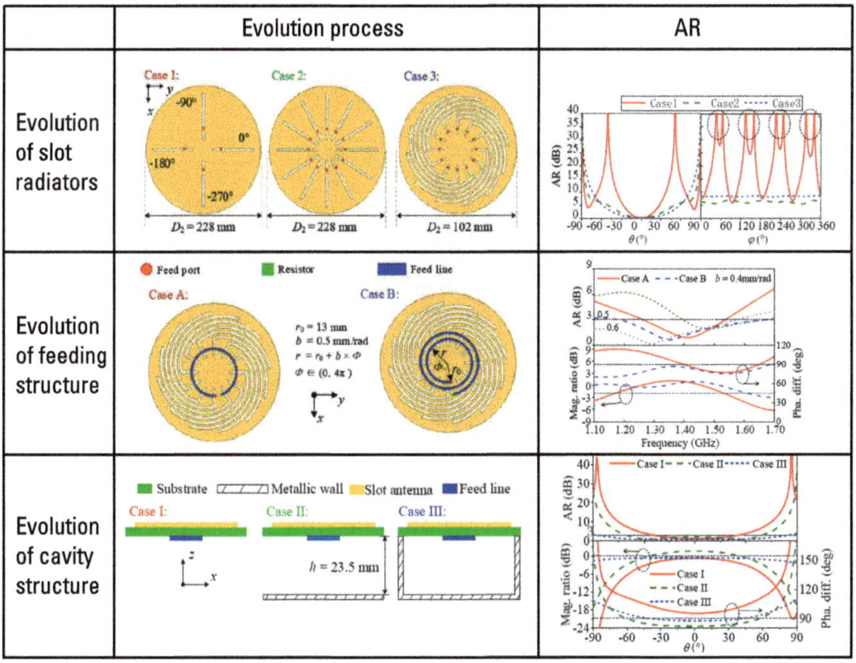

Figure 6.9 Evolution process of the antenna structure.

the 3-dB AR range achieved is limited to a specific bandwidth. To enhance AR bandwidth, a two-turn feeding line is implemented in case B. Its innovative design ensures a uniform magnitude and phase response over a broad frequency range. The two-turn feeding line comprises an inner round and an outer round, each aimed at improving the CP radiation at the lower and upper frequency bands. As the growth rate of the two-turn feed line b increases, the electrical length of the feeding line becomes longer, causing the optimal AR response to shift toward a lower frequency band.

Finally, the cavity structure is optimized in evolution. Case I in Figure 6.9 displays the side view of the aforementioned antenna without a reflector. While the bidirectional radiation pattern of this slot antenna is unsuitable for practical navigation terminal devices, a metallic reflector-backed, unidirectional version of Case II is designed to widen the 3-dB AR beamwidth from 66° to 118° at 1.4 GHz in the $\phi = 90°$ plane. The mechanism behind this improvement is the alteration of the phase response through field reflection from the reflector, resulting in a phase difference closer to 90° over a wider angular range. To further improve AR beamwidth, a modified cavity is designed in Case III. Strong currents are induced on the outside wall of the cavity, affecting the radiation of E_θ and E_ϕ differently at low elevation angles. The angular dependence of the magnitude ratio and phase difference between the two E-field components is significantly reduced, approximately equaling 0 dB and 90°, respectively. Increasing h_2 causes a decline in the resonant frequency of the spiral slots and feeding line, resulting in the movement of both AR and RHCP gain curves toward lower frequency bands. Raising the cavity diameter D_1 improves radiation efficiency of the proposed antenna, leading to higher RHCP gain within the GNSS band.

To validate the performance, an antenna prototype in Figure 6.8 was fabricated and tested, and the results are shown in Figure 6.10. The measured $|S_{11}|$ values below −10 dB span a frequency range of 1.15 to 1.7 GHz, which aligns closely with the simulated range of 1.12 to 1.7 GHz. The measured and simulated AR and RHCP gain exhibit a strong correlation across the frequency range, indicating excellent CP performance throughout all GNSS bands. At a frequency of 1.4 GHz, the realized AR beamwidths are impressively wide, reaching 144° and 180° in the planes of $\phi = 0°$ and $\phi = 90°$, respectively, and the measured 3-dB AR beamwidths at 1.2 and 1.6 GHz also exceed 140°.

A traveling-wave-fed slot spiral antenna with wide AR bandwidth and beamwidth was designed, analyzed, and validated. The arrangement of 12 spiral slots ensures AR uniformity in the elevation plane, and the implementation of a well-designed two-turn feed line enables the radiation of a broad AR bandwidth. A modified metallic cavity positioned beneath the radiating slots contributes to the attainment of a wide AR beamwidth. The proposed antenna exhibits satisfactory performance. The measured $|S_{11}|$ values cover the entire GNSS bands, and the antenna showcases wide 3-dB AR bandwidth and AR beamwidth.

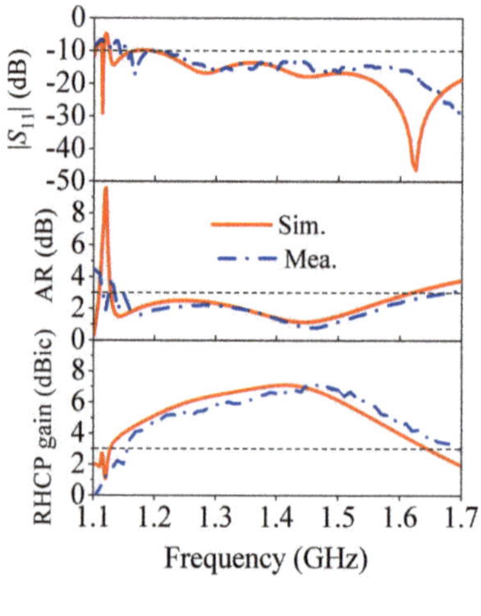

Figure 6.10 Simulated and measured results of the antenna: $|S_{11}|$, AR, and RHCP gain versus frequency.

Previous works have been proposed and developed using the multimode concept shown in Figure 6.11. In [4], the TM_{010} mode and ZOR mode of the antenna could be combined by using two radiators, gaining the wide E-plane HPBW of about 115°. In [9], the antenna resonates two modes by using the shorting pin and slot; in particular it realizes the E-plane beamwidth. In [10], the MPA operates at $TM_{3/4,1}$ and $TM_{9/4,1}$ modes while exhibiting a unidirectional beam with a wide half-power beamwidth of about 120° in its E-plane pattern.

6.3 Low Sidelobe Level Technique

This part discusses sidelobe level reduction of the antenna at E-plane or H-plane patterns.

6.3.1 Low H-Plane Sidelobe Level

In this part, we discuss design methods for reducing the sidelobe level of an antenna at H-plane. Figure 6.12 plots the detail configuration of a dual-band MPA with enhanced HPBW and low H-plane sidelobe level. It is composed of two radiation patches, shorting pins, dielectric substrate, feeding probes, and slot. Dual working bands are independently resonated and derived from the primary TM_{10} and TM_{12} modes. Meanwhile, the radiation pattern of these dual modes should be significantly improved and reshaped as shown in Figure 6.13. The E-plane HPBW of the antenna at improved TM_{10} mode could be widened, and the radiation pattern of the antenna at improved TM_{12} mode could be reshaped as the broadside beam with low H-plane sidelobe level. The specific parameters of the antenna are: $L = 45.9$ mm, $L_1 = 24.2$ mm, $L_2 = 19.7$

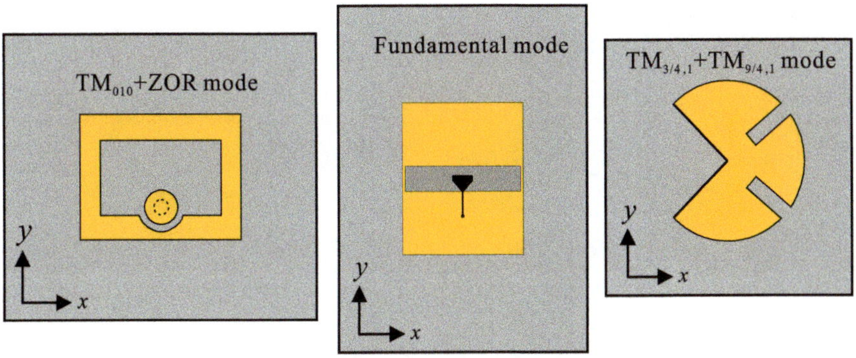

[4] [9] [10]

Figure 6.11 Previous antennas with wide beamwidth using the multimode-inspired concept.

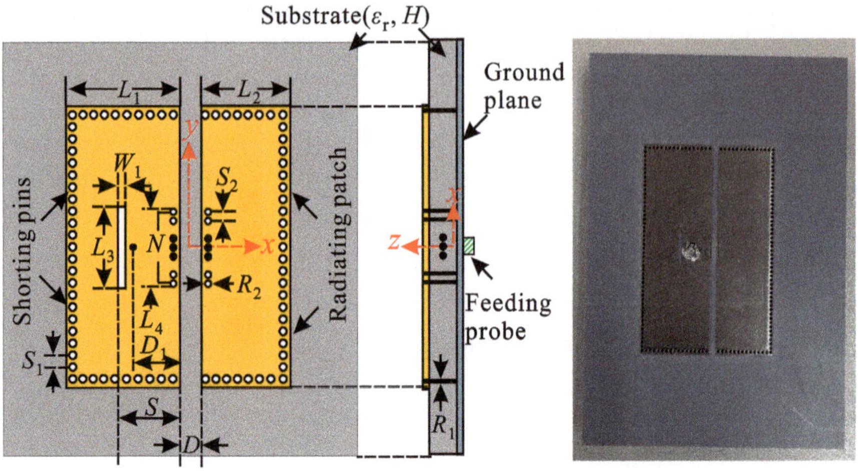

Figure 6.12 Schematic diagram and configuration of a dual-band multimode resonance-inspired MPA with enhanced E-plane HPBW and low H-plane sidelobe level.

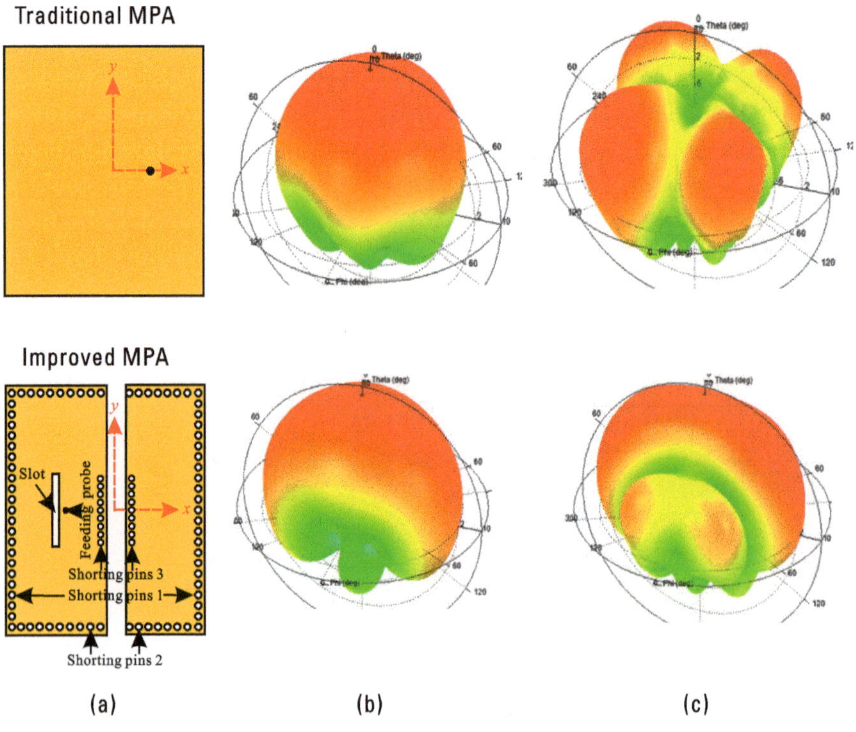

Figure 6.13 Radiation pattern comparison between a traditional MPA and the designed MPA in Figure 6.12: (a) configuration, (b) 3D-view pattern at TM_{10} mode, and (c) 3D-view pattern at TM_{12} mode.

mm, $L_3 = 11$ mm, $L_4 = 5.2$ mm, $L_g = 90$ mm, $W = 69.4$ mm, $W_1 = 1$ mm, $W_g = 130$ mm, $R_1 = 1$ mm, $R_2 = 0.4$ mm, $D = 2$ mm, $D_1 = 8.1$ mm, $S = 12.4$ mm, $S_1 = 1.4$ mm, $S_2 = 0.65$ mm, $N = 9$ mm. The relative dielectric constant of the substrate is 2.2 and the thickness is 4 mm [11].

A working principle for achieving wide E-plane HPBW at the lower band is discussed in Section 6.2.1. Here, the resonant mode is derived from the primary TM_{10} mode. We will illustrate the pattern improvement of the antenna with shorting pins along the y-direction at TM_{12} mode. Based on the cavity model, the *E*-plane radiation pattern of the shorted MPA in Figure 6.14(a) (antenna 1, which is derived from the Figure 6.2(b)) at improved TM_{12} mode could be expressed as

$$\begin{cases} E_\theta = 0 \\ E_\varphi = 0 \end{cases} \tag{6.7}$$

Meanwhile, Figure 6.14(a) shows the equivalent magnetic current and internal electric-field distribution of the antenna 1. The far-zone-radiated fields are mainly derived from the red equivalent magnetic currents (M_{s2}) and the green equivalent magnetic currents (M_{s1}). Unfortunately, they keep the same amplitude with opposite direction, producing small radiated fields in the broadside direction.

To address this issue, shorting pins are inserted around the edges of the core radiator along x-direction, named as antenna 2. Figure 6.14(b) plots the

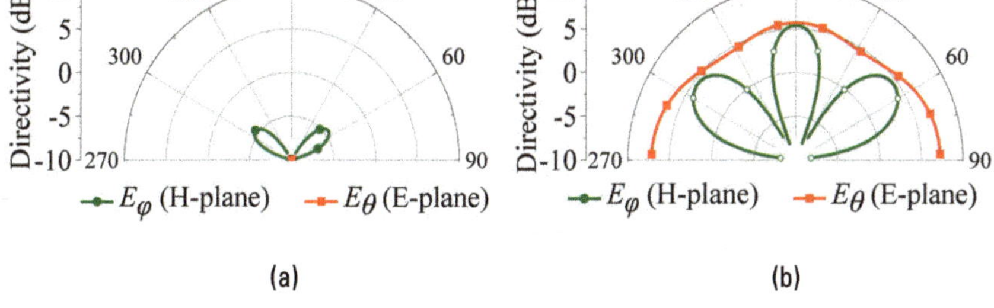

Figure 6.14 Schematic diagram of equivalent magnetic current, internal field, and radiation pattern comparison between antenna 1 and antenna 2 under reshaped TM_{12} mode: (a) antenna 1, and (b) antenna 2.

schematic diagram of the equivalent magnetic current, internal electric field, and far-zone-radiation pattern. Due to the introduced pins, the amplitude ratio between green equivalent magnetic currents (M_{s1}) and red equivalent magnetic currents (M_{s2}) could be reduced. The internal electric-field distribution of the improved TM_{12} mode resonance confirms this conclusion. With this improvement, the radiation null of the previous antenna 1 in the broadside direction is successfully reshaped into the broadside direction as desired. Most important, the *E*-plane HPBW of the antenna is widened to around 180° due to the single equivalent magnetic current, which is proved in the previous Section 6.2.1.

The results in Figure 6.14 show that the high H-plane sidelobe level is produced from the E_φ radiation pattern, which is mainly derived from the magnitude ratio between M_{s1} and M_{s2}. To control this magnitude ratio, shorting pins are inserted around M_{s1}, which is kept around the center of the radiating patch in Figure 6.15 (antenna 3). When the shorting pins are loaded at the center of the core radiator, the amplitude of its green equivalent magnetic currents M_{s1} is weakened, reducing the amplitude ratio n between M_{s1} and M_{s2}. The internal field distribution of antenna 3 under reshaped TM_{12} mode confirms this conclusion.

Figure 6.15 also depicts *H*-plane pattern variation under reshaped TM_{12} mode as a function of shorting pin number *N*. As *N* gradually increases, the amplitude of green equivalent magnetic currents $\overrightarrow{M}_{s1}$ gradually decreases, reducing the sidelobe level at H-plane radiation pattern. When *N* reaches to about 9, the sidelobe level of its *H*-plane radiation pattern is successfully reduced to around –9 dB.

The proposed multimode resonant antenna in Figure 6.12 was fabricated and measured. Figure 6.16 plots simulated and measured *S*-parameters, gain,

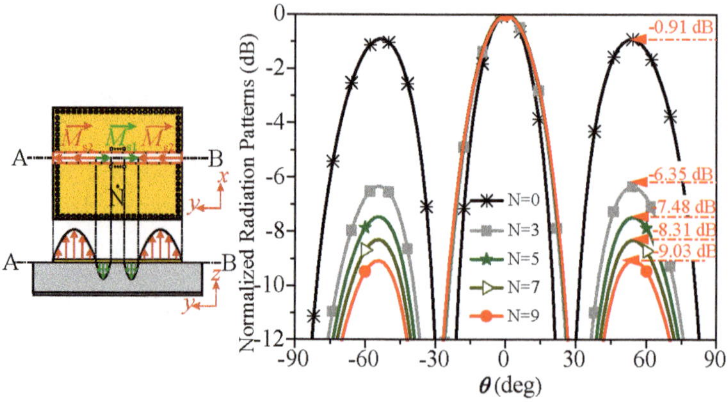

Figure 6.15 *H*-plane sidelobe level reduction of the antenna at TM_{12} mode under different number *N* of shorting pins.

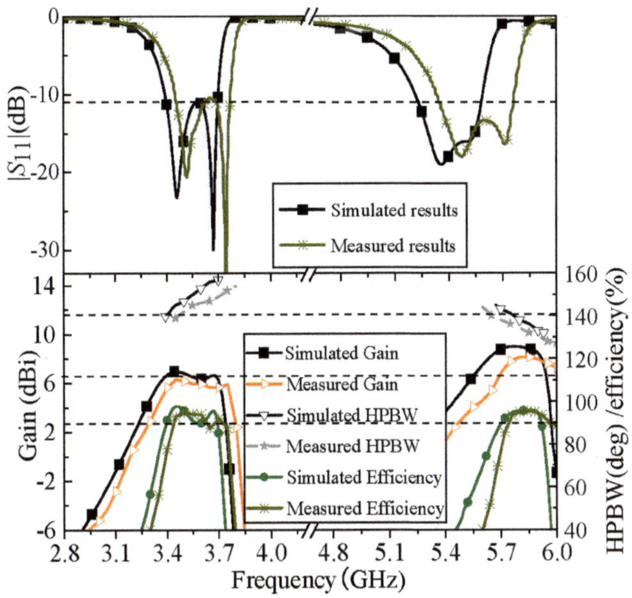

Figure 6.16 Simulated and measured $|S_{11}|$, gain, HPBW, and efficiency of the proposed dual-band multimode resonance-inspired antenna.

HPBW, and efficiency of the antenna as function of frequency, which show that the antenna has the desired impedance matching characteristics in dual working bands covering 3.45 to 3.77 GHz, and 5.75 to 6.04 GHz. In the lower band, the realized gain is kept around 6 dBi, making its E-plane HPBW around 150° with total efficiency near 90%. In the upper band, the antenna gain is maintained around 8 dBi, which is higher than that of the lower band due to increased aperture size. Meanwhile, the E-plane HPBW of the MPA is kept around 140° with total efficiency near 90%. The *H*-plane sidelobe level is successfully moved to below –10 dB over the upper working band as desired.

The HPBW of the antenna at the lower band is widened by using a single equivalent magnetic current instead of the traditional two equivalent magnetic currents. The nonbroadside beam of its TM_{12} mode is transformed into the broadside beam by loading pins at the patch edges along x-direction. The H-plane sidelobe of its reshaped TM_{12} mode antenna is reduced to control the magnitude ratio between two opposite equivalent magnetic currents, which could be realized by loading the shorting pins at the center of the radiating patch. The proposed multimode resonant antenna achieves low profile, dual wideband, stable gain, wide E-plane HPBW, and low H-plane sidelobe level characteristics simultaneously.

6.3.2 Low E-Plane Sidelobe and Adjustable H-Plane Beamwidth

Controlling E-plane and H-plane sidelobe levels simultaneously presents a significant challenge. This part introduces the working principle for realizing suppressed E-plane sidelobe and adjustable H-plane beamwidth. Figure 6.17 plots the evolution process of the proposed antenna, which is selected to be a square patch with a width of W= 60.0 mm, and slots loaded along the central line of patch in the x-direction in three different cases. The slot-loaded technique is developed on rectangular patch antennas operating under TM_{03} mode. In case 1, only a linear slot with the length of l_{s1} is introduced to the center of patch. In case 2, two-sided open slots are inserted from the two y-edges toward the center of patch with length of $l_{s2}/2$. And in case 3, both a central slot with length of l_{s1} and two-sided slots with length of $l_{s2}/2$ are simultaneously loaded. The slot widths are all set as W_s= 1 mm for the sake of low-fabrication error, and the patch is fed with a probe in a distance d to the rear-radiating edge. The substrate Rogers 5880 with a relative permittivity of 2.2, a loss tangent of 0.002, a thickness of 1.575 mm, and a copper surface roughness of 2.8 μm is employed [12].

First, we consider a conventional square patch antenna operating under its TM_{03} mode as shown in Figure 6.18. Using the cavity model [6], the surface current density on the patch can be mathematically expressed as a sine or cosine function with 3/2 periodicities in the y-direction while remaining constant in the x-direction, assuming the fringing effect around the four sides of the patch is negligibly small. It exhibits a radiation pattern in the E-plane similar to that of a 1.5-wavelength dipole.

As Figure 6.18 shows, etching a narrow slot along the central line of the rectangular patch in the x-direction forces out-of-phase current to circulate around it. This configuration divides the current density on the patch in two distinct parts that exhibit geometrically odd symmetry with respect to the H-plane. Due to their proximity, the far-field radiation produced by these two current parts almost cancels each other, resulting in minimal radiation. The SLL of

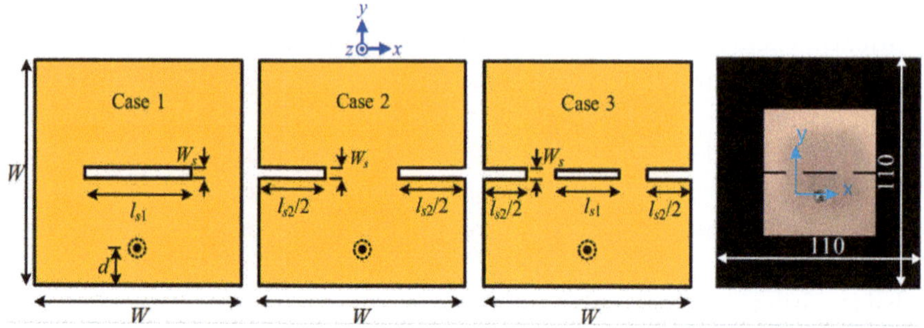

Figure 6.17 Geometries of three distinctive square patches with transversely loaded slots along the central line in the x-direction.

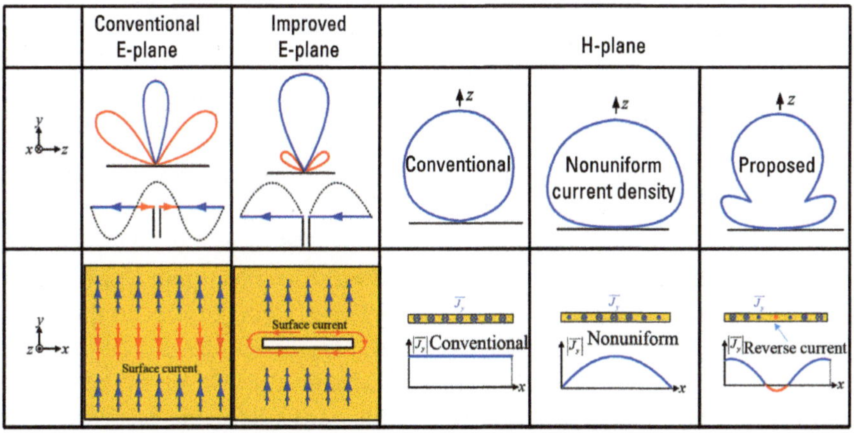

Figure 6.18 Surface current distribution and radiation patterns in E- and H-plane of TM_{03} mode.

the antenna in the E-plane experiences a significant reduction. This modification causes the effective current on the patch to behave similarly to that of a full-wavelength dipole, enhancing the antenna's radiation gain.

The distribution of y-oriented-current density J_y along the transverse section of the patch in the x-direction plays a crucial role in determining the beam shape and beamwidth of radiation patterns in the H-plane, as illustrated in Figure 6.18. In a conventional patch antenna, J_y is assumed to be constant in the x-direction, disregarding any edging effects, resulting in a relatively narrow beamwidth. However, by reshaping the distribution of J_y to be nonuniform, with its density decreasing from the center toward the edges of the patch, the beamwidth in the H-plane can be broadened to some extent. Conversely, if the current density decays from the edges inward and even becomes out of phase at the center, the radiation beam becomes more directive, potentially giving rise to sidelobes. By modifying the distribution of J_y, the beamwidth in the H-plane can be flexibly adjusted to meet specific requirements.

Based on the working principle in Figure 6.18, the three-slot-loading schemes in Figure 6.17 are employed to reshape the on-patch current distribution of the TM_{03} mode. All three slot configurations can suppress the E-plane sidelobe, since they can interrupt the out-of-phase current. However, they have different effects on the H-plane beamwidth because the transverse current distribution on the patch will be different in these three cases. More detailed results and discussion can be found in our reported works [12].

First, a central slot is introduced as shown in case 1 of Figure 6.17. The out-of-phase current in the patch center is interrupted to flow around the slot. As the slot length l_{s1}/W is increased, its radiative portion in the left and right sides will weaken and shrink. The SLL in the E-plane is steadily reduced but the HPBW in the H-plane is dramatically widened.

Second, the patch antenna loaded with two-sided slots is investigated, as shown in case 2 of Figure 6.17. In contrast to case 1, as slot length is increased, the transverse current density on the patch becomes stronger at the two sides and weaker in the center. Accordingly, the H-plane beamwidth is significantly narrowed and sidelobes come out.

An antenna combining the two above methods is proposed, as shown in case 3 of Figure 6.17. In this case, the transverse on-patch current density distribution can be manipulated by changing the length ratio of the central and side slots to achieve adjustable H-plane beamwidth as required. By choosing suitable total length and length ratio of three cascaded slots, the sidelobe in the E-plane can be reduced to its minimum and the beamwidth in H-plane narrowed at the same time. In this way, directivity can be effectively enhanced, which is higher than that of conventional patch antennas.

To validate the proposed method, the patch antenna with $l_{s1}/W = 0.3$ and $l_{s2}/W = 0.4$, was fabricated and measured. Geometric parameters are provided in Figure 6.17. The size of ground plane is selected as 110 mm × 110 mm in consideration of a tradeoff in backside radiation and fabrication cost. The simulated and measured reflection coefficients of the antenna are below –10 dB from 4.160 and 4.178 GHz to 4.224 and 4.238 GHz, respectively. Its simulated and measured radiation patterns in the two principal planes are shown in Figure 6.19. The main beam in the E-plane is narrowed with radiation nulls at around $\theta = \pm 55°$. Simulated and measured levels of sidelobes are below –20.8 and –18.6 dB, respectively. Beamwidth in the H-plane becomes quite narrow,

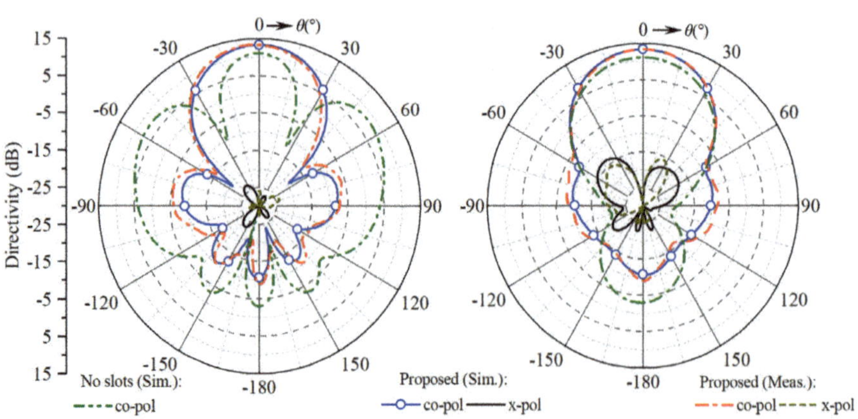

Figure 6.19 Simulated and measured results: (a) *E*-plane and (b) *H*-plane radiation patterns of the proposed antenna (at 4.200 GHz) and its counterpart without any slots (at 4.860 GHz) under TM_{03}-mode operation.

so the proposed antenna exhibits directivity as high as 13.3 dB. This directivity is increased by an extra 0.7 dB and the SLL is reduced by an extra 8.0 dB compared with the infinite ground case in simulation as analyzed.

The introduction of transverse slots along the central line of the patch effectively suppresses the sidelobes in the E-plane pattern of the antenna. Simultaneously, the sidelobes and beamwidth in H-plane pattern can be freely adjusted through various configurations of these slots. Such antennas maintain the advantage of high gain and satisfy signal coverage requirements for various applications.

Previous works have been proposed and developed by using the multimode concept as shown in Figure 6.20. In [13], the high-gain dual-mode patch antenna is designed based on the stacked disc radiator. Its fundamental working principle is based on the superposition of radiated far fields of the TM_{11} and TM_{13} modes. In [14], a set of slots are cut in the radiator to suppress E-plane and H-plane patterns at TM_{50}-like mode. As reported in [15], slots including pi, triangular, and circular slots are proposed to reduce the SLL at higher-order mode for the patch antenna.

6.4 Low Cross-Polarization Technique

Except for the main radiation beam, cross polarization is a key performance characteristic for the antenna. In this part, we discuss the working principle and design method for cross-polarization reduction: the shorting pins technique in Section 6.4.1, open-ended-stubs techniques in Section 6.4.2, and the slots-reconfiguration technique in Section 6.4.3.

6.4.1 Shorting Pins

Patch antennas with the loading of a pair of shorting pins are proposed toward flexible impedance matching and low cross polarization. Here, a dielectric sub-

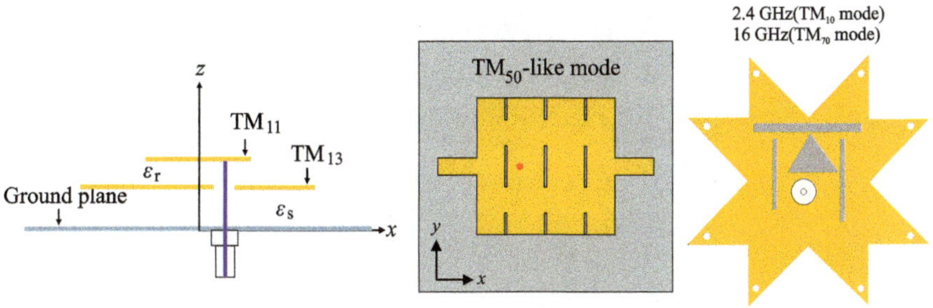

[13] [14] [15]

Figure 6.20 The previous antennas with high gain by using the multimode-inspired concept.

strate RO4003C with a relative permittivity of $\varepsilon_r = 3.55$, a thickness of $h = 1.524$ mm, a loss tangent of 0.0027, and a copper surface roughness of 2.8 μm is employed as default [16].

The geometry of the proposed antenna is depicted in Figure 6.21. Initially, the patch antenna, which is loaded with a single pin and operating in the TM_{01} mode, is fed by a microstrip line at the center of the radiating edge. A shorting pin is positioned along the centerline AA', with its distance to the feeding point set as *d*. The presence of the pin creates a shorting effect, enhancing the surface current density in its proximity. As the pin is brought closer to

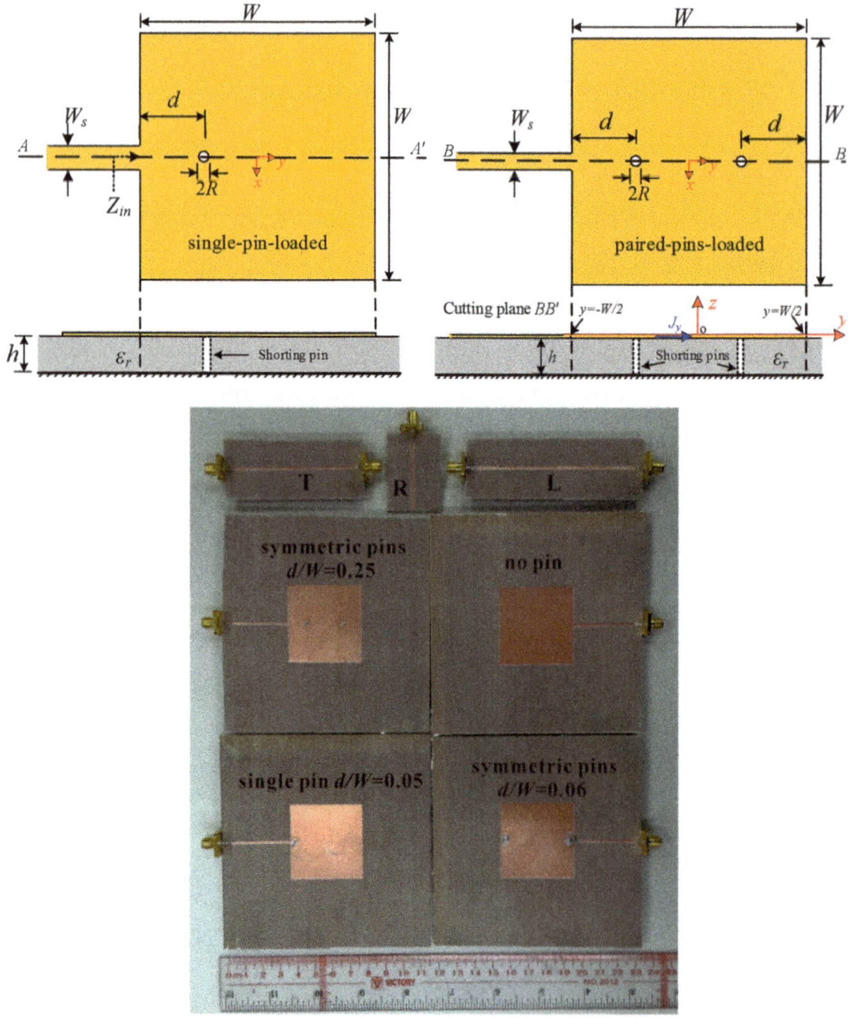

Figure 6.21 Geometry and configuration of the proposed antenna.

the feeding point, the current at the feeding edge increases, reducing the input impedance of the microstrip-edge-fed patch antenna.

The magnitude of y-oriented surface current density $|J_y|$ along the centerline is calculated and depicted in Figure 6.22. Placing the shorting pin in the center of the patch does not significantly affect the current density, which remains similar to that of a conventional square patch. However, as the pin is moved toward the feeding edge, the current density near the pin is disrupted. The surface current no longer exhibits an odd symmetric distribution with respect to the H-plane or *x-z* plane. As a result, the cancellation of far-zone electric fields is compromised, leading to an increase in cross polarization in the H-plane.

Figure 6.22 depicts the simulated radiation patterns of the patch antennas. The movement of the shorting pin from the center to the edge has a noticeable impact on the cross-polarization level in the H-plane. Specifically, near the two end-fire directions ($\theta = \pm 90°$), the cross-polarization level increases from –17.7 to –6.7 and –4.2 dBi when the pin location d/W is changed from 0.1 to 0.3 and 0.5, respectively. This observation indicates that the asymmetric placement of a single pin beneath the square patch significantly degrades the cross polarization in the H-plane.

To address the cross-polarization issue described above, a pair of shorting pins is symmetrically loaded on a square patch antenna along the centerline; its

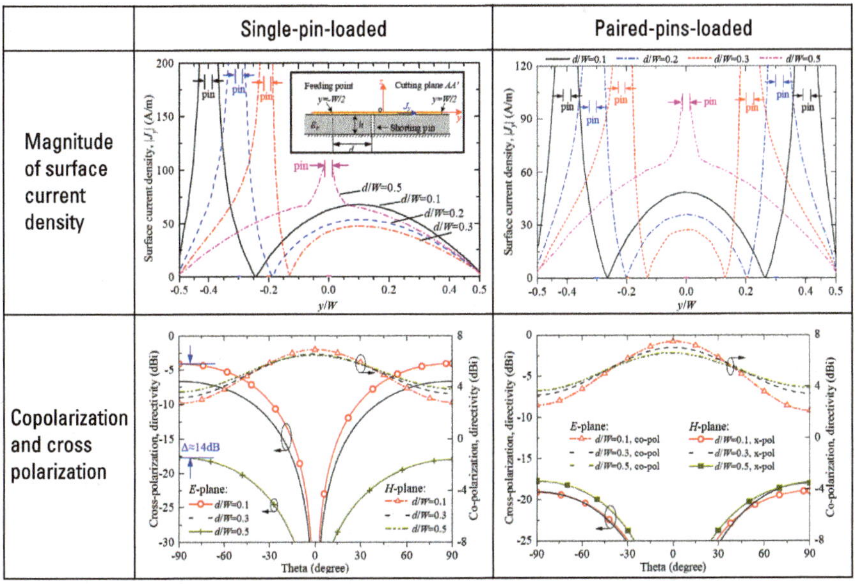

Figure 6.22 Magnitude of surface current density $|J_y|$ on the patch along the centerline and copolarization in *E*-plane, cross polarization in *H*-plane under different loaded pins, with infinite ground plane at different resonant frequencies, where $d/W = 0.1$, 0.3, and 0.5, respectively.

geometry is depicted in Figure 6.21. The patch is situated on the *x*-*y* plane, and the two shorting pins on the centerline BB' are symmetrically positioned with respect to the *x*-*z* plane. The size of the patch remains unchanged, ensuring a fair comparison between the modified antenna and the original one.

The magnitude of J_y under $d/W = 0.1$, 0.2, 0.3, and 0.5 is shown in Figure 6.22. Similar to the single-pin-loaded case, the current density at the feeding point ($y/W = -0.5$) is significantly enhanced, reducing the resonant input impedance of the edge-fed patch antenna. Additionally, the balanced loading arrangement with two pins ensures that the current density distribution $|J_y|$ becomes nearly symmetric along the direction from the front to the rear edge. When an extremely narrow feeding line is chosen, copolarization in the E-plane remains symmetric. The surface current is progressively distorted due to the shorting effect near these two pins. In contrast to the single-pin-loaded scenario discussed earlier, the two identical pins in this case are symmetrically positioned along the centerline relative to the center of the patch. The *x*-oriented current components exhibit odd symmetry with respect to the *x*-*z* plane. This cancellation of far fields radiated by these odd-symmetric *x*-oriented current vectors in the H-plane prevents a significant increase in cross polarization as observed in the single-pin-loaded patch antenna.

Figure 6.22 also shows the simulated radiation patterns of this antenna with $d/W = 0.1$, 0.3, and 0.5, or pin-to-pin spacing of 0.8 *W*, 0.4 *W*, and 0, respectively. Regardless of spacing between these paired pins, cross polarization in the H-plane is kept almost unchanged, and all are lower than –17.7 dBi. Compared to the –4.2 dBi cross-polarization directivity for the single-pin-loaded patch antenna, this design provides near 14-dB improvement.

For comparison, two paired-pins-loaded patch antennas with $d/W = 0.06$ and 0.25, and a conventional one without pin, were fabricated and measured as shown in Figure 6.21 to verify the theoretically predicted flexible impedance characteristics. Figure 6.23 depicts the radiation patterns in the E- and H-plane. They are well matched. Figure 6.23(a) shows that the maximum simulated and measured cross-polarization directivity of the single-pin-loaded patch antenna are –6.5 and –4.2 dBi, respectively. Instead, as shown in Figure 6.23(b), the cross-polarization directivity of the proposed paired-pins-loaded patch antenna is about –10 dB lower, with the maximum simulated and measured values of –16 and –14.5 dBi, respectively.

Symmetrically loading a pair of shorting pins beneath a square patch can strengthen surface current density near the pins. Input impedance at resonance can be adjusted over a wide range by changing the spacing between these pins. At the same time, the symmetric arrangement of these two identical pins makes the surface current on the patch odd symmetric with respect to the H-plane, keeping low cross polarization as expected.

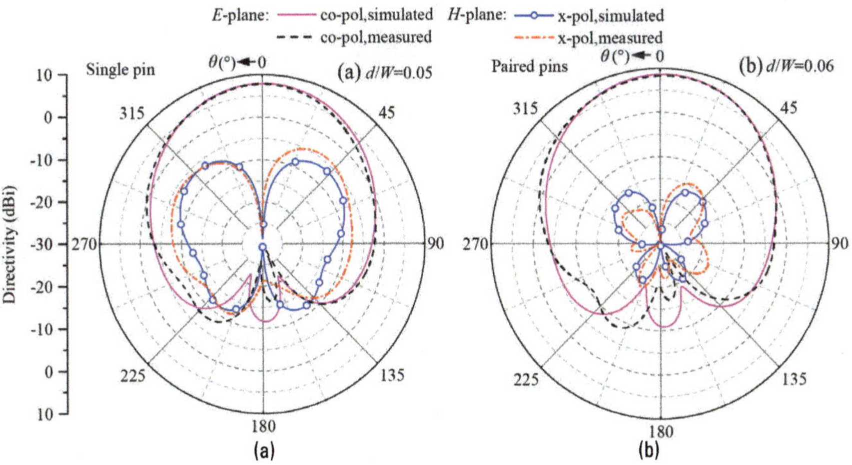

Figure 6.23 Simulated and measured radiation patterns of the two proposed patch antennas on a finite ground plane of 115 mm × 115 mm: (a) single-pin-loaded patch antenna with d/W = 0.05 at a resonant frequency of 2.241 GHz, and (b) paired-pins-loaded patch antenna with d/W = 0.06 at a resonant frequency of 2.478 GHz.

6.4.2 Open-Ended Stubs

As mentioned in Chapter 2, because the structure of an SPA is asymmetric at H-plane, it suffers from serious cross polarization at H-plane compared to a traditional MPA. To address this issue, open-ended stubs are loaded on the edges of the radiating patch. The detail structure diagram of the antenna with its fabrication is plotted in Figure 6.24, which is composed of a radiating patch, shorting pins, open stubs, and dielectric substrate. The $TM_{1/2,2}$ mode of the SPA is used here for cross-polarization reduction. The specific parameters of the antenna are: L_g = 150 mm, L_1 = 40 mm, L_2 = 4 mm, L_3 = 19.2 mm, W_g = 150 mm, W_1 = 26 mm, W_2 = 0.3 mm, R_1 = 1 mm, D = 20.5 mm. The relative dielectric constant of the dielectric substrate is 2.1; the thickness is 3 mm [17].

First, the schematic diagram of the internal electric field and equivalent magnetic current distribution of the traditional SPA at $TM_{1/2,2}$ mode are plotted in Figure 6.25. The equivalent magnetic currents are mainly composed of blue equivalent magnetic currents M_{s3} along the x-axis direction, red equivalent magnetic currents M_{s2} along the left, along the x-axis direction, and green equivalent magnetic currents M_{s1} along the y-axis direction. Under this condition, internal electric-field distribution from the M_{s1} and M_{s2} along the y-axis could be expressed as

$$E_z = -E_0 \cos(2\pi y / L_1) \tag{6.8}$$

where L_1 is the equivalent length of the radiation patch.

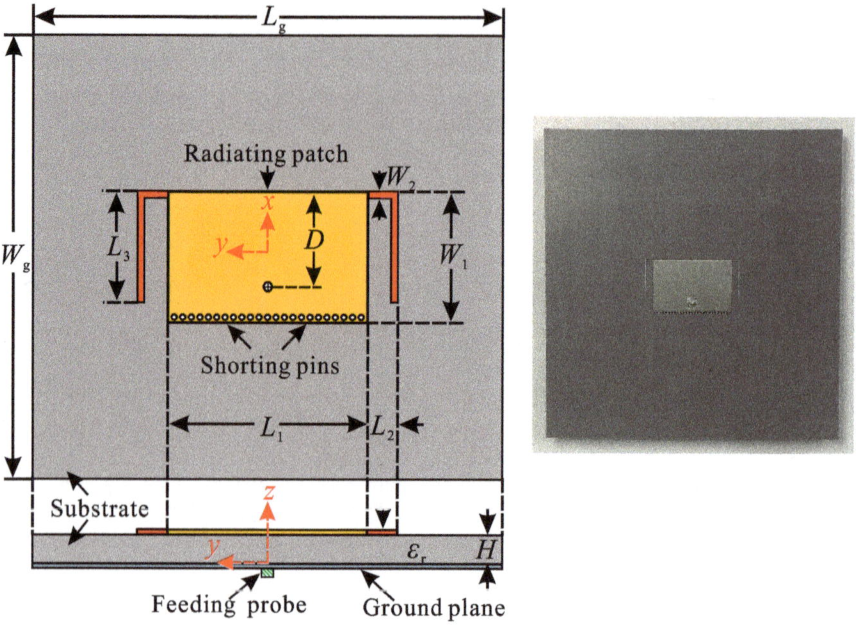

Figure 6.24 Structural diagram and fabrication of the proposed SPA with low H-plane cross polarization based on the open-ended stubs.

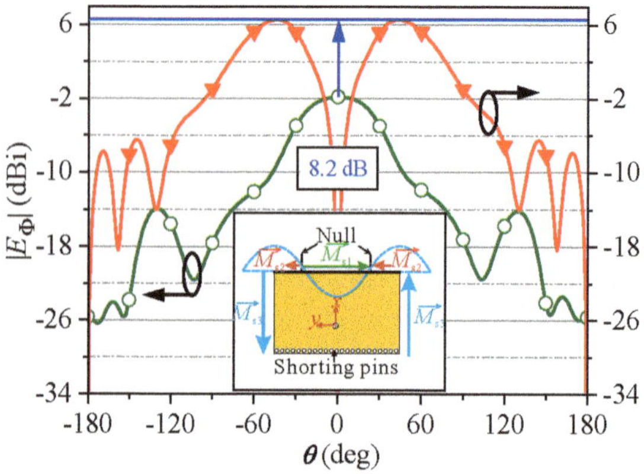

Figure 6.25 Schematic diagram of the internal field and equivalent magnetic current distribution for a traditional SPA under operation of $TM_{1/2,2}$ mode.

By using (6.8), its far-zone $|E_\theta|$ and $|E_\phi|$ components at *yoz*-plane could be written as

$$E_\theta = 0 \tag{6.9}$$

$$E_\varphi = jk_0 \frac{e^{-jkr}}{4\pi r} 2HE_0 \cos(\theta) \frac{L_1^2 \sin(\theta)/(\lambda\pi)}{1-(L_1 \sin(\theta)/\lambda\pi)^2} \sin(\frac{\pi L_1 \sin(\theta)}{\lambda}) \tag{6.10}$$

The internal electric fields from the M_{s3} along the x-axis could be expressed as

$$E_z = E_0 \sin\left[\pi x/(2W_1)\right] \tag{6.11}$$

where W_1 is the equivalent width of the radiation patch.

By using (6.11), far-zone $|E_\theta|$ and $|E_\phi|$ components at yoz-plane could be calculated as

$$E_\theta = jk_0 \frac{e^{-jkr}}{4\pi r} \left\{ j\frac{8HE_0W_1}{\pi} \sin\left[\frac{\pi L_1}{\lambda} \sin(\theta)\right]\right\} \tag{6.12}$$

$$E_\varphi = 0 \tag{6.13}$$

Based on (6.10) and (6.12), the equivalent magnetic currents along y-axis generate the $|E_\phi|$ component at yoz-plane, maintaining the normal radiation pattern, and the equivalent magnetic currents along the x-axis generate $|E_\theta|$ component at yoz-plane, maintaining a nonbroadside radiation pattern. To verify this predicted performance, Figure 6.25 also depicts the simulated radiation pattern of $TM_{1/2,2}$ mode at yoz-plane, which shows that the $|E_\theta|$ component keeps the peak gain off broadside direction with a large amplitude, where the $|E_\phi|$ component keeps the broadside beam with a small amplitude. The cross-polarized $|E_\theta|$ component is extremely high, requiring suppression.

Next, to realize low cross polarization, the amplitude ratio between M_{s1} and M_{s2} in Figure 6.25 is set as n, so the internal field function of (6.8) should be modified as

$$E_z = -E_0 \cos\left[2\pi y/(nL_1)\right] \tag{6.14}$$

By using (6.14), far-zone $|E_\theta|$ and $|E_\phi|$ components of the improved equivalent magnetic currents at yoz-plane could be represented as

$$E_\theta = 0 \tag{6.15}$$

$$E_{\varphi} = \cos(\theta)\frac{nL_l/\pi}{1-[nL_l\sin(\theta)/\lambda]^2}\left\{\sin(\frac{\pi}{n})\cos\left[\frac{\pi L_l}{\lambda}\sin(\theta)\right] - [nL_l\sin(\theta)/\lambda]\cos(\frac{\pi}{n})\sin\left[\frac{\pi L_l}{\lambda}\sin(\theta)\right]\right\} \quad (6.16)$$

From (6.16), the magnitude of $|E_\phi|$ component is mainly determined by the ratio *n*. Figure 6.26 discusses the $|E_\phi|$ component variation of $TM_{1/2,2}$ mode as a function of *n*. As *n* equal to 1, the $|E_\phi|$ component of the antenna exhibits a deep null in the broadside direction. By increasing the value of *n*, the $|E_\phi|$ component of the antenna in the normal direction gradually increases. To realize the largest $|E_\phi|$ value, *n* should be as large as possible. The magnitude ratio *n* could be controlled by setting the open-stub length.

As shown in Figure 6.27, $|E_\theta|$ and $|E_\phi|$ components of the SPA vary significantly under different stub lengths L_2. Initially, as L_2 equals 17.4 mm, the peak magnitude of $|E_\theta|$ component is much higher than that of $|E_\phi|$ component. By gradually enlarging the length L_2, the magnitude of $|E_\theta|$ component is reduced, whereas the magnitude of $|E_\phi|$ component is gradually increased. As L_2 equals 27.4 mm, the desired broadside radiation beam is successfully gained, while keeping a low cross-polarization level of around –20.1 dB. If L_2 continues to increase, the H-plane cross-polarization level of the antenna will be reduced to about –16.8 dB. To obtain the lowest cross-polarization level, the stub length L_2 should be 27.4 mm. Linear stubs are folded for size reduction.

The proposed antenna in Figure 6.24 was manufactured and tested. The designed antenna was successfully operated around 2.63 GHz. Figure 6.28

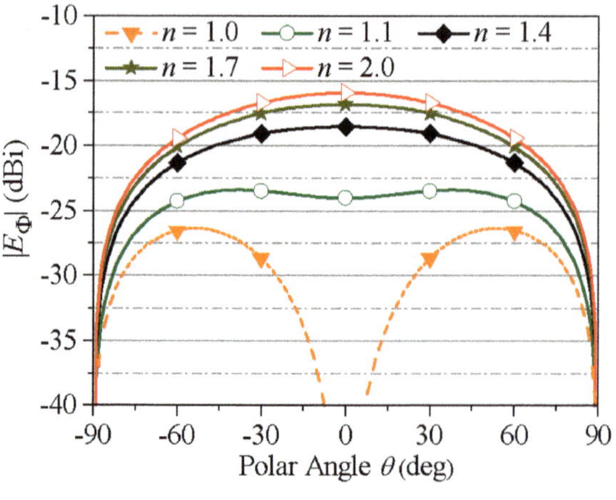

Figure 6.26 Calculated *yoz*-plane $|E_\phi|$ component of the SPA under radiation of $TM_{1/2,2}$ mode as a function of magnitude ratio *n*.

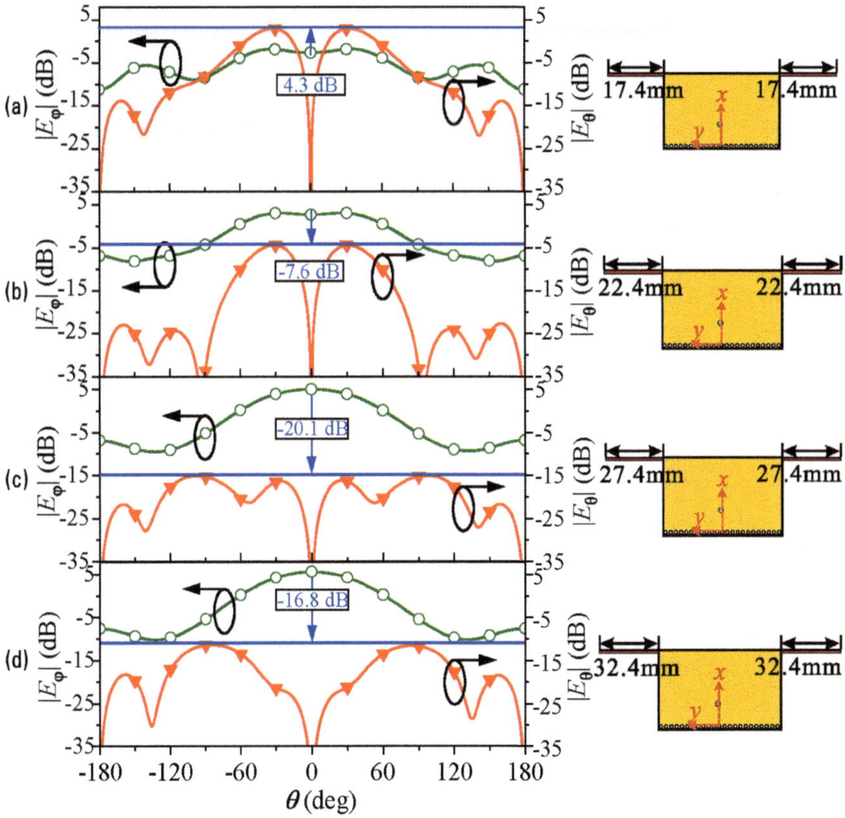

Figure 6.27 Radiation pattern variation of the stub-loaded SPA loaded with different lengths L_2: (a) $L_2 = 17.4$ mm, (b) $L_2 = 22.4$ mm, (c) $L_2 = 27.4$ mm, and (d) $L_2 = 32.4$ mm.

plots the simulation and test radiation pattern of the antenna and the traditional SPA for radiation pattern comparison. Compared to a traditional SPA, the *H*-plane cross polarization of this SPA is reduced to about –16.1 dB. Internal fields between a traditional SPA and our proposed stub-loaded SPA are also compared in Figure 6.28. When a traditional SPA is operated at $TM_{1/2,0}$ mode, its maximum internal field is distributed around the upper corners of the radiating patch, as shown in Figure 6.28(a), resulting in a high H-plane cross-polarization level. In contrast, when the folded stubs are loaded on the SPA, as shown in Figure 6.28(b), the internal field around the corners of the radiating patch is significantly weakened, resulting in low H-plane cross-polarization level characteristics as desired.

In summary, open-ended stubs are used to tune internal field nodes of $TM_{1/2,2}$ mode around the upper edges of the radiation patch, achieving a reduction in H-plane cross-polarization level. The proposed multimode

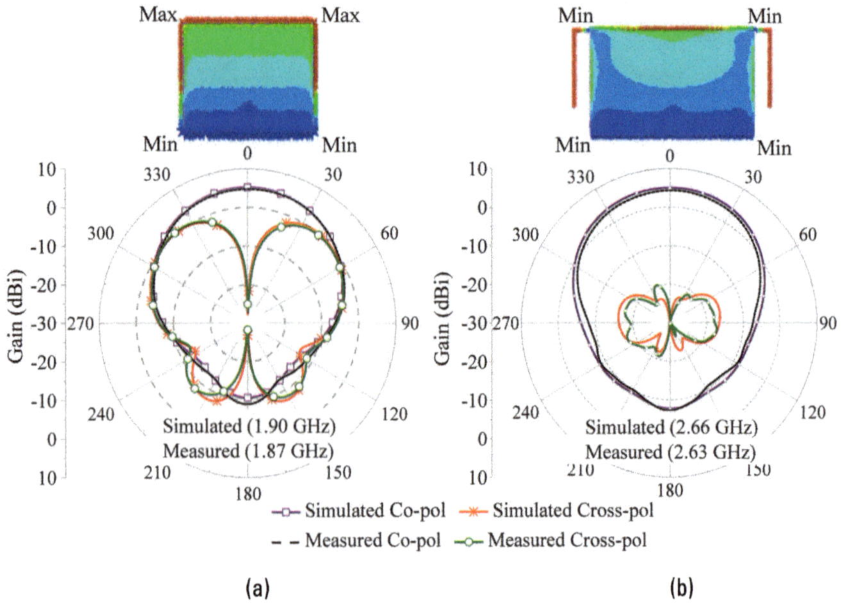

Figure 6.28 Radiation pattern comparison between two types of SPA at H-plane: (a) traditional SPA, and (b) proposed SPA.

resonance-inspired antenna has low profile, miniaturization, high efficiency and low H-plane cross-polarization characteristics simultaneously.

6.4.3 PIN Diodes Loaded and Differentially Fed Multimode Resonant MPA

In this section, a slot-loaded high-gain circular MPA is proposed to realize reconfigurable ±45° polarizations and low cross polarization. The configuration of the proposed antenna is illustrated in Figure 6.29, a circular patch loaded with two *y*-oriented slots 1, two *x*-oriented slots 2, four fan-shaped slots 3, and four DC lines printed on the top layer of a substrate ($\varepsilon_r = 3.55$ and $H_{sub} = 0.814$ mm). Another DC line is printed on the bottom layer of the substrate and connected to center of the patch with a shorting pin. The substrate is suspended above a ground plane with a distance of H_1 by four plastic supports. Two probes differentially feed the patch from the backside of the ground. Geometrical parameter values are: $R_g = 80$ mm, $R_p = 44$ mm, $R_o = 23.8$ mm, $R_i = 22$ mm, $W_{s1} = 2.5$ mm, $W_{s2} = 2$ mm, $L_{dc1} = 34$ mm, $L_s = 22$ mm, $D_f = 41.2$ mm, $H_1 = 4$ mm, $H_{sub} = 0.813$ mm. More details about the antennas can be found in our reported works [18].

Figure 6.30 shows the evolution of the circular patch illustrating three types of slot-loading schemes without PIN diodes. Initially, the current vectors on circular patch antenna of TM_{31} mode were circumferentially distributed

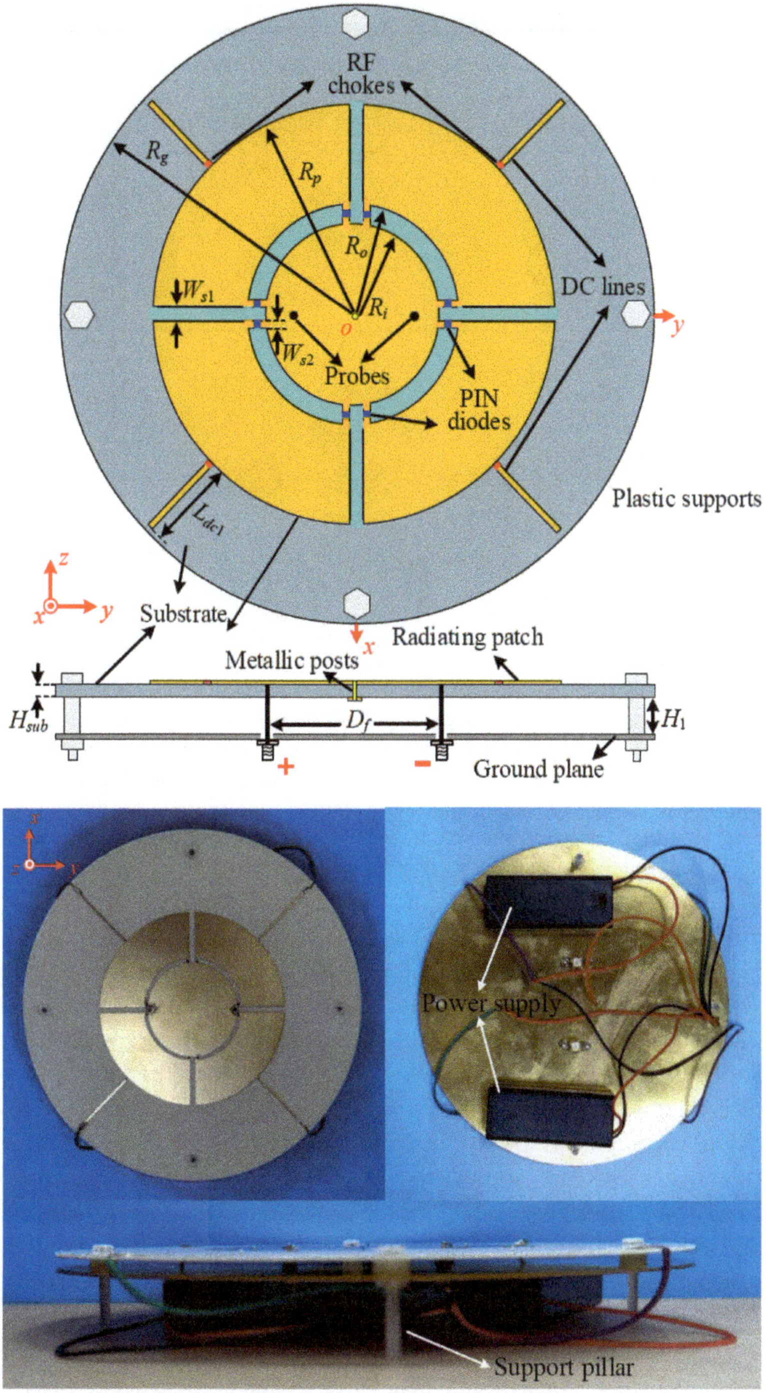

Figure 6.29 Geometry and configuration of the proposed high-gain antenna with low cross polarization.

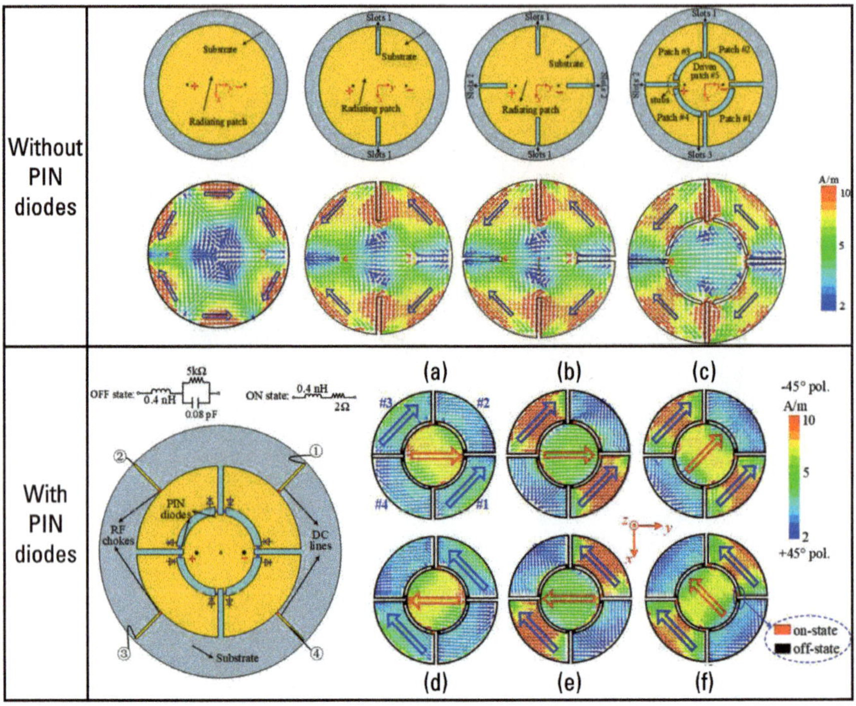

Figure 6.30 Evolution process of a slot-loaded MPA and current distributions of the proposed polarization-reconfigurable (PR) antenna under six types of connection topologies, the 45°-polarized states: (a) case 1 with two y-oriented strips; (b) case 2 with two x-oriented strips; (c) case 3 with four strips, and the –45°-polarized state; (d) case 4 with two y-oriented strips; (e) case 5 with two x-oriented strips; and (f) case 6 with four strips.

with six sinusoidal half-period, which led to a radiation null in the broadside direction and resulted in a conical radiation pattern. However, when slots 1 were introduced along the *y*-axis of the patch, they perturbed two segments of *x*-oriented currents, reducing the cancelling effect among the currents. As the length of slot 1 increases, the currents become more in-phase and the radiation of *x*-oriented currents becomes progressively dominant. The beam will transform from conical to normal, featuring enhanced broadside radiation.

This accomplishes the goal of achieving a highly directive beam. The next step involves introducing additional slot-loadings parallel to the currents of the modified TM_{31} modes and reconfiguring the patch with minimal perturbation. As illustrated in Figure 6.30, the current vectors near the *x*-oriented slots 2 were weak and parallel to slots 2, so there was no discernible impact on the impedance or radiation performance.

Subsequently, fan-shaped slots 3 were incorporated in the patch at the same radius. These slots, mostly parallel to the adjacent currents, had minimal impact on the current distribution of the patch, so the resonant frequency experienced only a slight shift. The radiation pattern of the antenna, with the inclusion of slots 2 and 3, maintains a directive beam at the broadside, similar to the case with only slots 1.

So far, the conventional circular patch has been divided into a smaller driven patch #5 and four fan-shaped patches #1, #2, #3, and #4, which are electrically connected by short strips at two ends of slots 3.

Analyzing the current distribution in the final case in Figure 6.30 without PIN diodes, the radiative currents on four fan-shaped patches are either +45° or –45° polarized. To realize reconfigurable ±45° polarizations with high directivity and low cross-polarization, the connection between the driven patch #5 and four fan-shaped patches is investigated by changing ON/OFF state of the connecting strips. The configuration of the proposed PR patch antenna using PIN diodes is also shown in Figure 6.30, and the surface current distribution of slot-loaded patch under six types of connection topologies is illustrated. Eight Bar50-02L PIN diodes provided by Infineon Technologies [19] are placed at two ends of each of slots 3 to control the connection between patches #1 to #4 and driven-patch #5. ON/OFF state is denoted by red/black segment and the direction of dominant current vectors is denoted with blue/red arrows.

In case 1, as illustrated, when the driven-patch #5 connects only to patches #1 and #3 through two conductive strips along the *y*-axis, the current vectors on the driven-patch #5 flow along the x-axis and are more robust than those on the fan-shaped patches flowing in the 45° direction. This configuration results in significant –45° cross polarization.

Conversely, in case 2, when the two connecting strips are oriented along the x-axis, the current density on patches #1 and #3 becomes much stronger, dominating the 45°-polarization radiation. However, effective currents on the driven-patch #5 are maintained in the *x*-direction, leading to cross polarization.

In case 3, both patches #1 and #3 are connected to the driven-patch #5 using four ON-state strips. In this configuration, the current vectors on each patch are reshaped to align in the 45° direction. This alignment enhances copolarization and reduces cross polarization.

Similarly, if –45° polarization is required, parasitic patches #2 and #4, as well as driven-patch #5, are connected under different configurations through PIN diodes. As shown in Figure 6.30(d, f), from cases 4 to 5 and 6, the –45°-polarized current is progressively enhanced, improving polarization purity.

Figure 6.29 shows a photograph of the proposed PR antenna. The DC lines and batteries are placed on the backside of the 0.5-mm-thick copper ground to avoid mutual interference with antenna radiation. Its reflection coefficients for both ±45°-polarization states are simulated and measured. The overlapped impedance bandwidth ($|S_{dd}| < -10$ dB) in simulation for both working states is 5.7% in a range from 2.37 to 2.51 GHz. The measured value shifted about 18 MHz lower than the simulated value due to manufacturing tolerances and unexpected strays of lumped elements. The radiation patterns of the proposed antenna in two principal planes at 2.44 GHz are plotted in Figure 6.31. A satisfactory agreement between the simulated and measured results is achieved. Switchable ±45° polarizations realize a high gain up to 10.3 dBi, whereas normalized cross polarization is suppressed below –25 dB and the front-to-back ratio is as high as 23 dB. This achieves high gain, low cross polarization, and low backside radiation.

The slot-loaded high-gain circular patch antenna proposed to realize reconfigurable ±45° polarizations is achieved. Different reconfigurable topologies are studied by dividing the patch into five segments through two sets of slots. By changing the connection states of different segments through PIN diodes, the switchable ±45° polarizations with high-polarization purity are realized with low cross polarization.

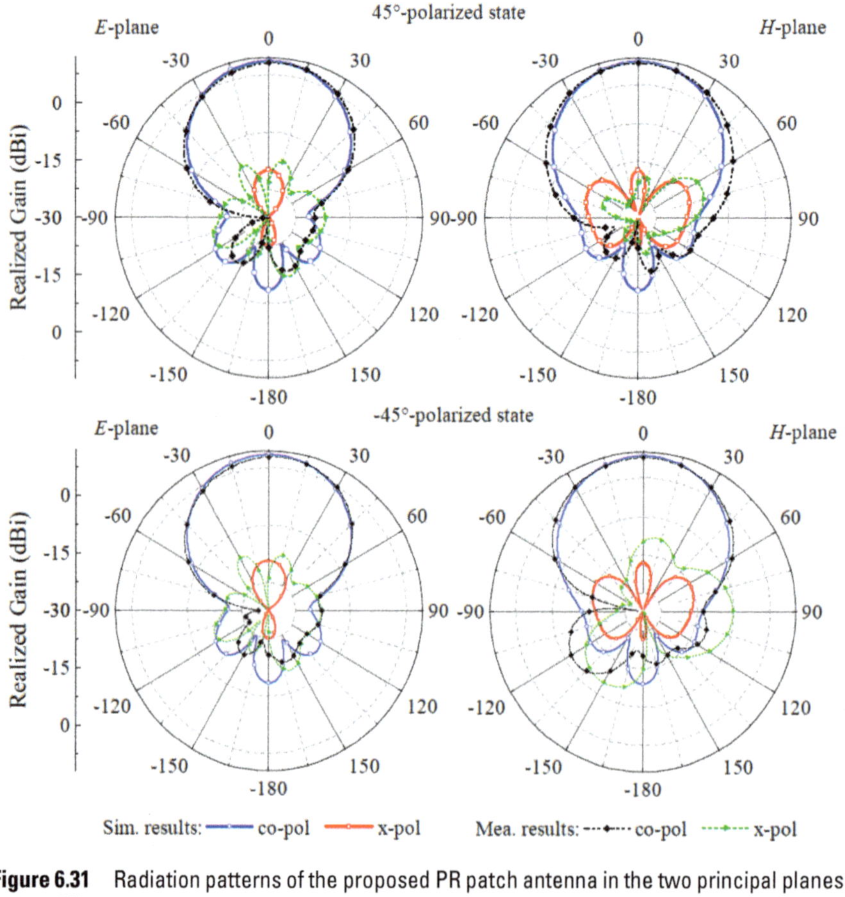

Figure 6.31 Radiation patterns of the proposed PR patch antenna in the two principal planes at 2.44 GHz.

Previous works have been proposed and developed using the parasitic elements, DGS, and MCM shown in Figure 6.32. In [20], parasitic elements are loaded to improve the cross-polarization discrimination of a broadband ±45° dual-polarized base station antenna within ±60° of the main lobe in the horizontal plane. In [21], a modified-cavity model (MCM) is demonstrated as a method for cross-polarization reduction.

6.5 Beam-Scanning Bandwidth Technique

Apart from the main beam in a fixed direction, the radiation beam of the antennas is also required with scanning performance over a wide operating band. In this part, we discuss the working principle and design method for beam scanning, including single beam scanning in Section 6.5.1, and dual-beam scanning in Section 6.5.2.

6.5.1 Single-Beam Scanning

In a traditional leaky-wave antenna, EH_1 mode is used to generate single-beam scanning performance [23]. This design is also large. To reduce the size, the lower EH_0 mode could be employed, but it keeps the dual-beam radiation pattern. Shaping dual beams of EH_0 mode into a broadside beam could generate single-beam scanning with wide bandwidth and a compact size. Figure 6.33 shows the structure of the leaky-wave antenna under EH_0 mode, which is mainly composed of radiating patches, stubs, slits, dielectric substrates, differential feeding, and ground plane. The specific parameters of the antenna are: $W_{sub}=$ 60 mm, $L_{sub}=$ 312.5 mm, $H_{sub}=$ 1.5 mm, $W_p=$ 4 mm, $L_p=$ 9.4 mm, $L_s=$ 7mm, $W_s=$ 0.5 mm, $L_{stub}=$ 5.08 mm, $W_c=$ 1 mm, $L_c=$ 8.6 mm, $L=$ 262.5 mm, $W_{slit}=$ 1 mm, $W_{stub}=$ 1 mm, $D_r=$ 2.72 mm, $x_1=$ 6 mm, $x_2=$ 3.4 mm, $W=$ 7 mm, $T=$

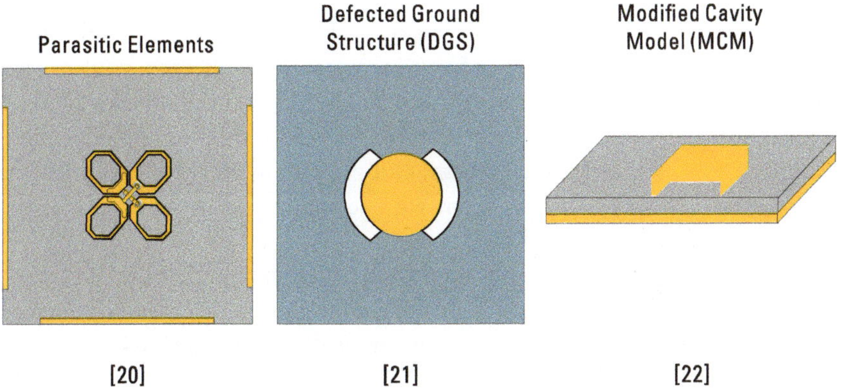

Figure 6.32 Previous antennas with cross-polarization reduction.

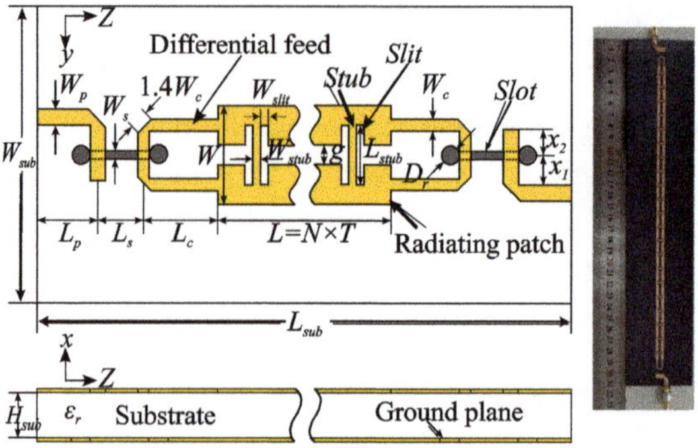

Figure 6.33 Configuration and fabrication of a single-beam leaky-wave antenna based on EH_0 mode.

12.5 mm, g= 2 mm, N= 21. The dielectric constant of the dielectric substrate is 2.55 and the thickness is 1.5 mm.

First, we discuss the evolution of a single-beam leaky-wave antenna based on the EH_0 mode, as shown in Figure 6.34. For the traditional leaky-wave antenna in Figure 6.34(a), the tangential electric field of its EH_0 mode around two side edges keeps the opposite direction, as illustrated in Figure 6.34(b), resulting in the undesired dual-beam pattern shown in Figure 6.34(c). Considering the image effect beneath the ground, the antenna in Figure 6.34(a) is rotated and constructed in Figure 6.34(d). The previous dual-beam pattern is also rotated 90°, forming up an 8-type directional pattern shown in Figure 6.34(e). After that, the differential feeding scheme is alternatively employed to produce the equivalent electric wall at the center of the patch. An additional ground plane is introduced underneath the overall antenna, which is plotted in Figure 6.34(f). Internal electric-field distributions keep the in-phase performance, which is plotted in Figure 6.34(g). With this arrangement, the single broadside beam is generated and reshaped in Figure 6.34(h) as desired.

The proposed antenna derived from the element in Figure 6.34(f) was designed, fabricated, and measured. The electrical length of the proposed antenna in Figure 6.33 is only about $0.155\lambda_e$ (λ_e represents the effective guided wavelength of the antenna at the lower band 4.5 GHz). Figure 6.35 shows the simulated and measured results of the proposed leaky-wave antenna, which contains of S-parameter, single-scanning beam, and gain simultaneously. It proves that the antenna generates a single beam with scanning performance in the wide operation band from 4.5 to 5.5 GHz.

Figure 6.35 also depicts the simulated and measured gain responses of a leaky-wave antenna based on EH_0 mode. It shows that as the working frequency increases, the H-plane beam-scanning angle of the designed antenna gradually decreases from 82° to 30°. Meanwhile, the maximum realized gain of the antenna reaches to about 12 dBic over the wide band. The designed leaky-wave

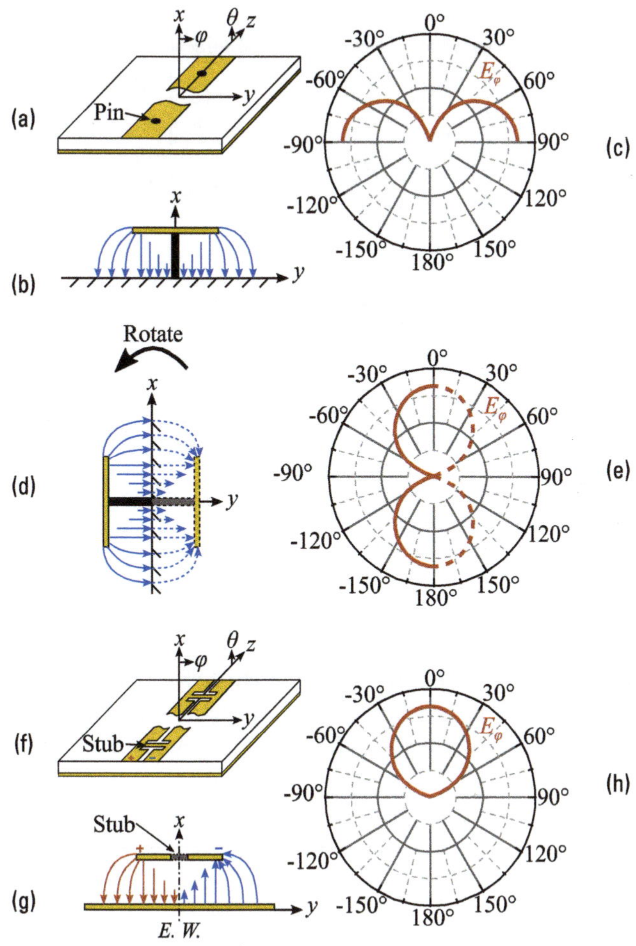

Figure 6.34 Evolution of the leaky-wave antenna from the dual beam to single beam via EH_0 mode: (a) traditional antenna, (b) internal electric-field distribution of a traditional antenna, (c) radiation pattern of a traditional antenna, (d) internal electric-field distribution of a rotated antenna, (e) 8-type radiation pattern of a rotated antenna, (f) designed antenna with single beam, (g) internal electric fields of the designed antenna with single beam, and (h) radiation pattern of the designed antenna.

antenna has the desired single-beam patterns in the entire band, while keeping low cross-polarization level characteristics.

The dual beam of a leaky-wave antenna at EH_0 mode is reshaped into single-beam radiation by rotating the antenna configuration and adding the differential feeding scheme. The proposed multimode resonance-inspired antenna achieves the desired characteristics of small size, low profile, high gain, and single-beam scanning.

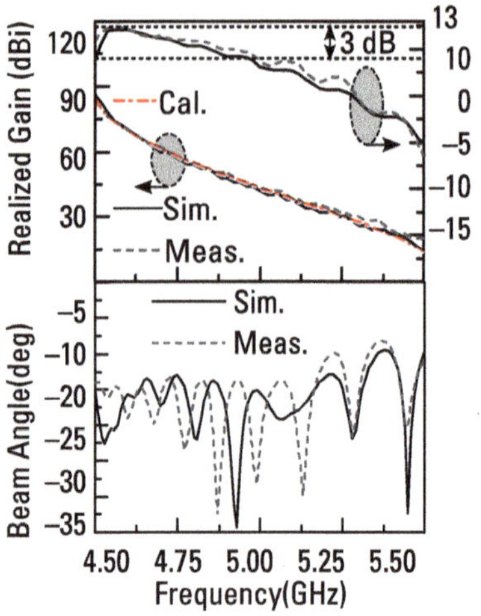

Figure 6.35 Simulated and measured S-parameters, scanning angle, and realized gain of a single-beam leaky-wave antenna based on EH_0 mode.

6.5.2 Dual-Beam Scanning

Dual-beam scanning performance is also valuable. This part discusses the working principle for achieving dual-beam performance in wideband with wide scanning angle. Figure 6.36 shows the configuration of a leaky-wave antenna based on EH_2 mode, which is composed of a radiating patch, slot, feeding network, dielectric substrate, and ground plane. The scanning angle of the dual-beam leaky-wave antenna is mainly determined by the open stopband during the wideband. The specific parameters of the antenna are: $W = 40$ mm, $W_s = 1$

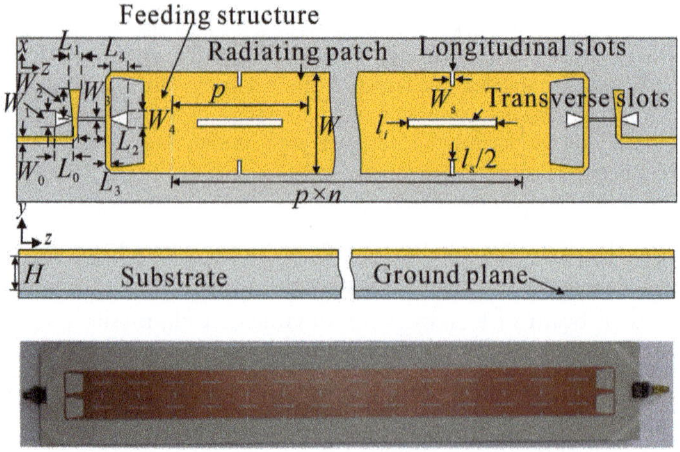

Figure 6.36 Configuration and fabrication of a proposed dual-beam leaky-wave antenna based on EH_2 mode.

mm, $W_0 = 2.5$ mm, $W_1 = 0.3$ mm, $W_2 = 3.2$ mm, $W_3 = 1.5$ mm, $W_4 = 3$ mm, $L_0 = 1.6$ mm, $L_1 = 10$ mm, $L_2 = 9$ mm, $l_s = 4.8$ mm, $l_i = 16.9$ mm, p = 30 mm, $H = 0.635$ mm, $n = 14$, the dielectric constant of the dielectric substrate is 6.15, and the thickness is 0.635 mm [24].

Figure 6.36 shows that the transverse and longitudinal slots are both cut on the radiating patch. We will investigate the internal electric-field distribution and equivalent circuit model of the antenna element based on the loaded slot, plotting the relevant results in Figure 6.37. The transverse slot introduces impedance Z_1, and the longitudinal slot introduces impedance Z_2.

Based on the above equivalent circuit model, Figure 6.38 gives the $Y_{sh}/2$, $Z_{se}/2$, and propagation constants of the antenna element as a function

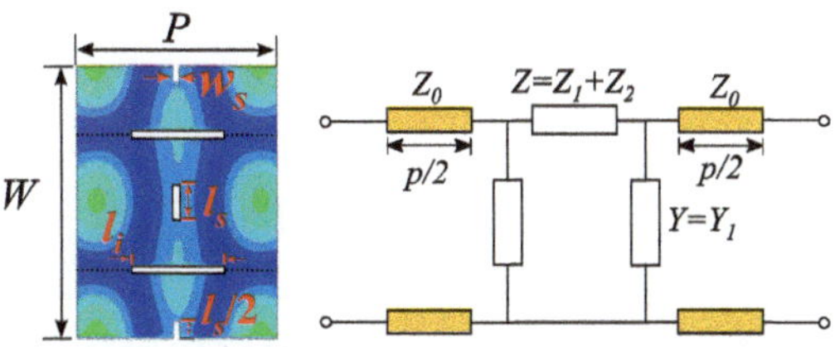

Figure 6.37 Internal electric-field distribution and equivalent circuit model of a leaky-wave-antenna element based on transverse slot and longitudinal slot.

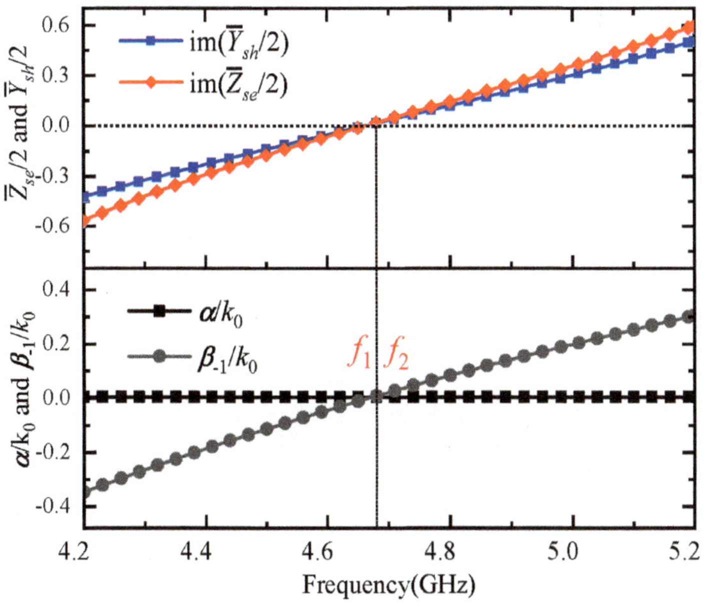

Figure 6.38 $Y_{sh}/2$, $Z_{se}/2$ propagation constant of the leaky-wave antenna as a function of frequency.

of frequency. The resonant frequencies f_1 and f_2 could be overlapped around 4.7 GHz, effectively suppressing the open stopband without a serious drop in the attenuation constant at broadside regime. By using short-open calibration technology [25], the propagation constant can be obtained to exhibit coupling between adjacent cells.

As shown in Figure 6.36, the proposed antenna was designed, fabricated, and measured. The designed antenna has good impedance matching characteristics in working frequencies from 3.6 to 6.0 GHz, inherent performance for a leaky-wave antenna. Figure 6.39 shows the simulated and measured far-zone-radiation patterns of the designed antenna based on EH_2 mode. When the operating frequency is reduced from 5.9 to 3.9 GHz, the scanning angle of the dual beams at H-plane could be increased as desired from $\theta = 49°$ to $\theta = 130°$ due to the suppressed open stopband.

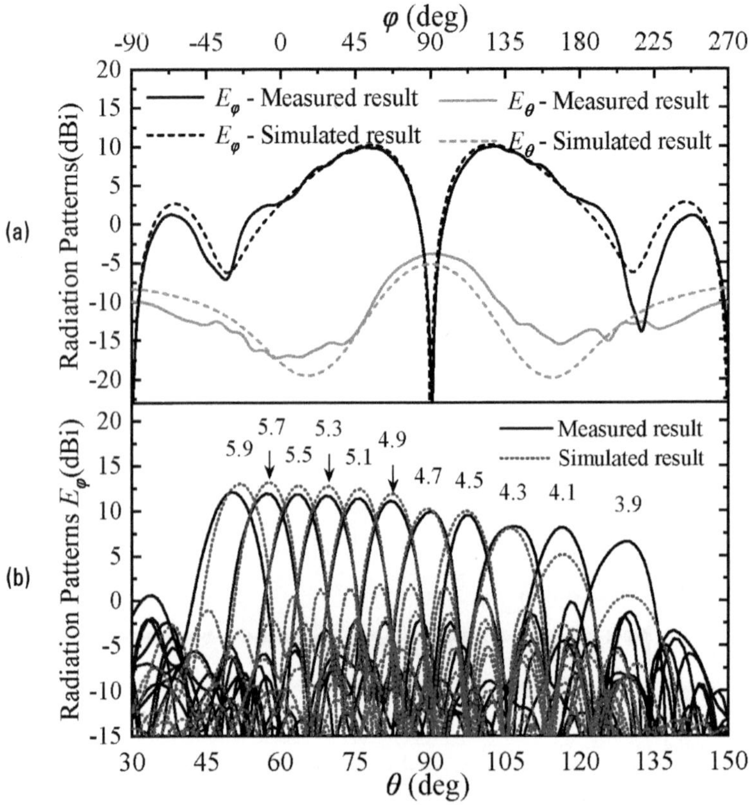

Figure 6.39 Simulated and measured radiation patterns of a leaky-wave antenna with dual-beam scanning at E-plane and H-plane based on EH_2 mode.

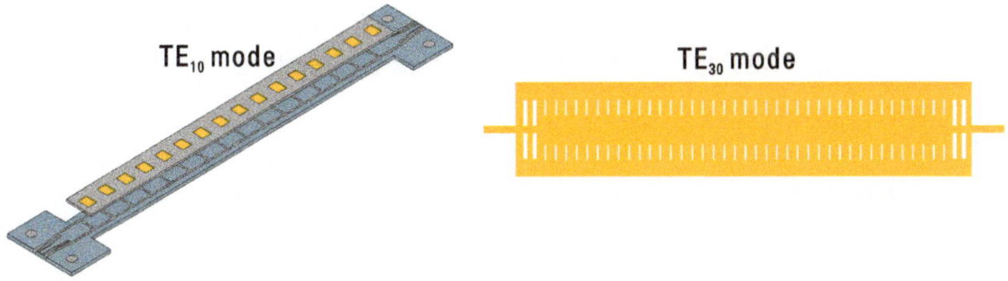

[26] [27]

Figure 6.40 Previous leaky-wave antennas with improved beam scanning.

The transverse slot and longitudinal slot are loaded to address the open-stop band defect for traditional leaky-wave antennas at EH_2 mode, realizing the characteristics of dual-beam scanning performance over the wide angle. The designed multimode resonance-inspired leaky-wave antenna achieves the desired low profile, small size, wide frequency band, dual beam, and large scanning angle from forward to backward directions. Previous works have been proposed and developed by using the multimode-inspired concept as shown in Figure 6.40. In [26], compared with TE_{10} mode, antennas at TE_{30} and TE_{50} mode have transverse beamwidths markedly narrowed, and gains improved by about 5.5 and 8 dB, respectively. In [27], a scheme based on the multimode resonator (MMR) concept is proposed and demonstrated for the design and development of leaky-wave antennas (LWAs) to realize stable radiation properties.

6.6 Conclusions

The radiation patterns of antennas can be enhanced through several approaches. For example, antenna HPBW could be widened by reducing aperture size, MPA cross-polarization levels could be suppressed by using DGSs, antenna sidelobe levels could be reduced by using arrays. Compared to these above-mentioned methods, however, mode-inspired antennas with pattern improvements shown in this section have the advantages of compact size, stable beam, or low cross-polarization level. These proposed antennas have many potential applications in radar communication, phased arrays, wearable systems, and more.

References

[1] Yang, W. -J., Y. -M. Pan, and S. -Y. Zheng, "A Low-Profile Wideband Circularly Polarized Crossed-Dipole Antenna with Wide Axial-Ratio and Gain Beamwidths," *IEEE Transactions on Antennas and Propagation*, Vol. 66, No. 7, July 2018, pp. 3346–3353.

[2] Yang, G., et al., "Broadening the Beam-Width of Microstrip Antenna by the Induced Vertical Currents," *IET Microwaves, Antennas & Propagation,* Vol. 12, No. 2, February 2018, pp. 190–194.

[3] Li, M., S. Xiao, and B. Wang, "Compact, Low-Profile, HIS-Based Pattern-Reconfigurable Antenna for Wide-Angle Scanning," *8th European Conference on Antennas and Propagation (EuCAP 2014),* September 2014, pp. 1541–1544.

[4] Ko, S. -T., and J. -H. Lee, "Hybrid Zeroth-Order Resonance Patch Antenna with Broad E-plane Beamwidth," *IEEE Transactions on Antennas and Propagation,* Vol. 61, No. 1, January 2013, pp. 19–25.

[5] Liu, N. -W., et al., "A Low-Profile Printed Cavity Antenna with Simultaneous Bandwidth and Radiation Pattern Improvement," *IEEE Antennas and Wireless Propagation Letters,* Vol. 18, No. 10, October 2019, pp. 2125–2129.

[6] Lo, Y., D. Solomon, and W. Richards, "Theory and Experiment on Microstrip Antennas," *IEEE Transactions on Antennas and Propagation,* Vol. 27, No. 2, March 1979, pp. 137–145.

[7] Liu, N. -W., et al., "A Low-Profile Dual-Band Patch Antenna with Simultaneous Wide Beamwidth and High Gain by Using Multiresonant Modes," *IEEE Antennas and Wireless Propagation Letters,* Vol. 20, No. 5, May 2021, pp. 813–817.

[8] Zhong, Z. P., and X. Zhang, "A Travelling-Wave-Fed Slot Spiral Antenna with Wide Axial-Ratio Bandwidth and Beamwidth for GNSS Applications," *IEEE Open Journal of Antennas and Propagation,* Vol. 2, 2021, pp. 578–584.

[9] Row, J. -S., and Y. -H. Chen, "Wideband Planar Array with Broad Beamwidth and Low Cross Polarization," *IEEE Transactions on Antennas and Propagation,* Vol. 63, No. 9, September 2015, pp. 4161–4165.

[10] Lu, W. -J., et al., "Generalized Design Approach to Compact Wideband Multiresonant Patch Antennas," *International Journal of RF Microwave Computer-Aided Engineering,* Vol. 28, September 2018.

[11] Liu, N. -W., et al., "Dual-Band Single-Layer Microstrip Patch Antenna with Enhanced Bandwidth and Beamwidth Based on Reshaped Multiresonant Modes," *IEEE Transactions on Antennas and Propagation,* Vol. 67, No. 11, November 2019, pp. 7127–7132.

[12] Zhang, X, L. Zhu, and Q. S. Wu, "Sidelobe-Reduced and Gain-Enhanced Square Patch Antennas with Adjustable Beamwidth Under TM_{03} Mode Operation," *IEEE Transactions on Antennas and Propagation,* Vol. 66, No. 4, April 2018, pp. 1704–1713.

[13] Juyal, P., and L. Shafai, "A High-Gain Single-Feed Dual-Mode Microstrip Disc Radiator," *IEEE Transactions on Antennas and Propagation,* Vol. 64, No. 6, June 2016, pp. 2115–2126.

[14] Santos, R. -A., et al., "Gain Enhancement and Sidelobe Level Reduction of Microstrip Patch Antenna Under Operation of TM_{50}-like Mode," *IEEE Open Journal of Antennas and Propagation,* Vol. 5, No. 4, August 2024, pp. 983–993.

[15] Khan, Q. U., D. Fazal, and M. bin Ihsan, "Use of Slots to Improve Performance of Patch in Terms of Gain and Sidelobes Reduction," *IEEE Antennas and Wireless Propagation Letters,* Vol. 14, October 2015, pp. 422–425.

[16] Zhang, X., and L. Zhu, "Patch Antennas with Loading of a Pair of Shorting Pins Toward Flexible Impedance Matching and Low Cross Polarization," *IEEE Transactions on Antennas and Propagation,* Vol. 64, No. 4, April 2016, pp. 1226–1233.

[17] Liu, N. -W., et al., "Cross-Polarization Reduction of a Shorted Patch Antenna with Broadside Radiation Using a Pair of Open-Ended Stubs," *IEEE Transactions on Antennas and Propagation,* Vol. 68, No. 1, January 2020, pp. 13–20.

[18] Hong, K. -D., et al., "A Slot-Loaded High-Gain Circular Patch Antenna with Reconfigurable Orthogonal Polarizations and Low Cross Polarization," *IEEE Antennas and Wireless Propagation Letters,* Vol. 21, No. 3, March 2022, pp. 511–515.

[19] Data Sheet of Bar50-02L PIN Diodes, Infineon Technologies, Application Note, September 2017, http://www.infineon.com/.

[20] Luo, Y., Q. -X. Chu, and D. -L. Wen, "A Plus/Minus 45 Degree Dual-Polarized Base-Station Antenna with Enhanced Cross-Polarization Discrimination Via Addition of Four Parasitic Elements Placed in a Square Contour," *IEEE Transactions on Antennas and Propagation,* Vol. 64, No. 4, April 2016, pp. 1514–1519.

[21] Guha, D., C. Kumar, and S. Pal, "Improved Cross-Polarization Characteristics of Circular Microstrip Antenna Employing Arc-Shaped Defected Ground Structure (DGS)," *IEEE Antennas and Wireless Propagation Letters,* Vol. 8, December 2009, pp. 1367–1369.

[22] Ghosh, D., et al., "Physical and Quantitative Analysis of Compact Rectangular Microstrip Antenna with Shorted Nonradiating Edges for Reduced Cross-Polarized Radiation Using Modified Cavity Model," *IEEE Antennas and Propagation Magazine,* Vol. 56, No. 4, August 2014, pp. 61–72.

[23] Duan, J., and L. Zhu, "An EH_0-Mode Microstrip Leaky-Wave Antenna on Differentially Excited Coupled Line for Transversal Single-Beam Radiation," *IEEE Transactions on Antennas and Propagation,* Vol. 69, No. 10, October 2021, pp. 6941–6946.

[24] Zhang, P. –F, L. Zhu, and S. Sun, "Microstrip-Line EH_1/EH_2-mode Leaky-Wave Antennas with Backward-to-Forward Scanning," *IEEE Antennas and Wireless Propagation Letters,* Vol. 19, No. 12, December 2020, pp. 2363–2367.

[25] Li, Y., and L. Zhu, "A Short–Open Calibration Method for Accurate De-embedding of 3-D Nonplanar Microstrip Line Structures in Finite-Element Method," *IEEE Transactions on Microwave Theory and Techniques,* Vol. 66, No. 3, March 2018, pp. 1172–1180.

[26] Liu, J., and J. Liang, "Gain Enhancement of Transversely Slotted Substrate Integrated Waveguide Leaky-Wave Antennas Based on Higher Modes," *IEEE Transactions on Antennas and Propagation,* Vol. 69, No. 8, August 2021, pp. 4423–4438.

[27] Zheng, D., and K. Wu, "Leaky-Wave Antenna Featuring Stable Radiation Based on Multimode Resonator (MMR) Concept," *IEEE Transactions on Antennas and Propagation,* Vol. 68, No. 3, March 2020, pp. 2016–2030.

About the Authors

Neng-Wu Liu was born in Changde, China. He received BS and ME degrees in electrical engineering from Xidian University, Xi'an, China, in 2012 and 2015, respectively, and a PhD degree from the University of Macau, Macau, in 2017.

From 2018 to 2024, he was an associate professor in the National Key Laboratory of Antennas and Microwave Technology at Xidian University, Xi'an, China. In 2024 he became a professor in the National Key Laboratory of Radar Detection and Sensing at Xidian University, Xi'an, China. From 2019 to 2021, he was the UM Macau Postdoctoral Research Fellow under the UM Macau Talent Program. His current research interests include the designs of antenna theory, low-profile antenna, multimode antenna, wideband antenna, patch antenna, dielectric resonator antenna, circularly polarized antennas, filtering antenna, slot antenna, and phased array.

Professor Liu has been the recipient of several academic awards, which includes the Best Student Paper award in the *17th IEEE Macau/HK AP/MTT* postgraduate conference, and Best Student Paper award in the *2017 National Conference on Antennas.* He was a recipient of an Outstanding Reviewer award from the *IEEE Transactions on Antennas and Propagation* in 2020–2024, Macau Natural Science awards (second prize) from the Science and Technology Development Fund (FDCT) Macau in 2022, and the Natural Science award of China Communications Society in 2022 and 2024 (first prize), and nomination for the Natural Science Award of the people's Republic of China in 2025. He serves as an associate editor of *IET Microwaves, Antennas & Propagation, IEEE Access,* and *IET Electronics Letters.*

Xiao Zhang was born in Gaozhou, Guangdong, China. He received a BEng in information engineering and a MEng in communication and information systems from the South China University of Technology, Guangzhou,

China, in 2011 and 2014, respectively, and a PhD in electrical and computer engineering from the University of Macau, Macau, SAR, China, in 2017.

From September 2012 to August 2014, he was a research assistant with Comba Telecom Systems Limited, Guangzhou, China. In 2018, he joined the College of Electronics and Information Engineering, Shenzhen University, Shenzhen, China, where he is currently an associate professor in the College of Electronics and Information Engineering. His research interests include high-gain antennas, wideband antennas, circularly polarized antennas, terminal antennas, filtering antennas, reflectarray, and characteristic mode analysis.

Dr. Zhang was the recipient of the Scientific and Technological R&D award for Postgraduates of Macau in 2018. He was recognized as the Shenzhen Overseas High Caliber Personnel Level C in 2018. He was also recognized in the World's Top 2% Scientists list in 2019, 2022, and 2024 by Elsevier.

Lei Zhu received BEng and MEng degrees in radio engineering from the Nanjing Institute of Technology (now Southeast University), Nanjing, China, in 1985 and 1988, respectively, and a PhD in electronic engineering from the University of Electro-Communications, Tokyo, Japan, in 1993.

From 1993 to 1996, he was a research engineer with Matsushita-Kotobuki Electronics Industries Ltd., Tokyo, Japan. From 1996 to 2000, he was a research fellow with the École Polytechnique de Montréal, Montréal, Quebec, Canada. From 2000 to 2013, he was an associate professor with the School of Electrical and Electronic Engineering, Nanyang Technological University, Singapore. He joined the Faculty of Science and Technology, University of Macau, Macau, China, as a full professor in 2013, was promoted as a Distinguished Professor in 2016, and then as a Chair Professor in 2023. From 2014 to 2017, he served as the head of the Department of Electrical and Computer Engineering, University of Macau. His research interests include microwave circuits, antennas, periodic structures, electromagnetic surfaces, and computational electromagnetics.

Dr. Zhu was an associate editor for the *IEEE Transactions on Microwave Theory and Techniques* (2010–2013) and *IEEE Microwave and Wireless Components Letters* (2006–2012). He served as a member of the IEEE MTT-S Fellow Evaluation Committee (2013–2015), and as a member of IEEE AP-S Fellows Committee (2015–2017). He was the recipient of the *1997 Asia-Pacific Microwave Prize Award*, the 2020/2022/2024 Macau Natural Science Awards (second prize), nomination for the Natural Science Award of the people's Republic of China 2025, and the *2024 IEEE MTT-S Microwave Application Award*. He was elevated to a IEEE Fellow in November 2011.

Index

Recent Titles in the Artech House Antennas and Propagation Library

Christos Christodoulou, Series Editor
Vince Rodriguez, Series Editor

Adaptive Array Measurements in Communications, M. A. Halim

Antenna-on-Chip: Design, Challenges, and Opportunities, Hammad M. Cheema, Fatima Khalid, and Atif Shamim

Antenna Design for Cognitive Radio, Youssef Tawk, Joseph Costantine, and Christos Christodoulou

Antenna Design for CubeSats, Reyhan Baktur

Antenna Design with Fiber Optics, A. Kumar

Antenna Engineering Using Physical Optics: Practical CAD Techniques and Software, Leo Diaz and Thomas Milligan

Antenna Measurement Techniques, Gary E. Evans

Antennas for IoT, Prutha P. Kulkarni

Antennas and Propagation for Body-Centric Wireless Communications, Second Edition, Peter S. Hall and Yang Hao, editors

Antennas and Site Engineering for Mobile Radio Networks, Bruno Delorme

Antennas for Small Mobile Terminals, Kyohei Fujimoto and Koichi Ito, editors

Analysis, Design and Measurement of Small and Low-Profile Antennas, Kazuhiro Hirasawa

Analysis of Radome-Enclosed Antennas, Second Edition, Dennis J. Kozakoff

Applications of Geographic Information Systems for Wireless Network Planning, Francisco Saez de Adana, Josefa Gómez Pérez, Abdelhamid Tayebi Tayebi, and Juan Casado Ballesteros

The Art and Science of Ultrawideband Antennas, Second Edition, Hans G. Schantz

AWAS for Windows Version 2.0: Analysis of Wire Antennas and Scatterers, Antonije R. Djordjević, et al.

Broadband Microstrip Antennas, Girsh Kumar and K. P. Ray

Broadband Patch Antennas, Jean-François Zürcher and Fred E. Gardiol

Circularly Polarized Dielectric Resonator Antennas, Raghvendra Kumar Chaudhary, Rajkishor Kumar, and Rakesh Chowdhury

Design and Applications of Active Integrated Antennas, Mohammad S. Sharawi and Oualid Hammi

Design and Measurement of Antennas and Propagation in Mobile Cellular Systems, Hiroyuki Arai

Designer Notes for Microwave Antennas, Richard C. Johnson

Dielectric Resonator Antenna Handbook, Aldo Petosa

Dielectric Resonator Antennas, Biswajeet Mukherjee and Monika Chauhan

Electromagnetics and Antenna Technology, Alan J. Fenn

Electromagnetics, Microwave Circuit and Antenna Design for Communications Engineering, Peter Russer

Engineering Applications of the Modulated Scatterer Technique, Jean-Charles Bolomey and Fred E. Gardiol

Fixed & Mobile Terminal Antennas, Akhileswar Kumar

Foundations of Antenna Engineering: A Unified Approach for Line-of-Sight and Multipath, Per-Simon Kildal

Four-Arm Spiral Antennas, Joseph A. Mosko

Frequency-Agile Antennas for Wireless Communications, Aldo Petosa

Fresnel Zones in Wireless Links, Zone Plate Lenses and Antennas, Hristo D. Hristov

Handbook of Antennas for EMC, Second Edition, Thereza Macnamara with John McAuley

Handbook of Reflector Antennas and Feed Systems, Volume I: Theory and Design of Reflectors, Satish Sharma, Sudhakar Rao, and Lotfollah Shafai, editors

Handbook of Reflector Antennas and Feed Systems, Volume II: Feed Systems, Lotfollah Shafai, Satish Sharma, and Sudhakar Rao, editors

Handbook of Reflector Antennas and Feed Systems, Volume III: Applications of Reflectors, Sudhakar Rao, Lotfollah Shafai, and Satish Sharma, editors

Interference Suppression Techniques for Microwave Antennas and Transmitters, Ernest Freeman

Introduction to Active Phased Arrays, Thomas Sikina

Introduction to Antenna Analysis Using EM Simulators, Hiroaki Kogure, Yoshie Kogure, and James C. Rautio

Introduction to Antennas and RF Propagation Analysis, Dean James Friesen

LONRS: Low-Noise Receiving Systems Performance and Analysis Toolkit, Charles T. Stelzried, Macgregor S. Reid, and Arthur J. Freiley

Mechanical and Manufacturing Design of Antennas, Richard L. Burge

Measurement of Mobile Antenna Systems, Second Edition, Hiroyuki Arai

Microstrip Lines and Slotlines, Third Edition, Ramesh Garg, Inder J. Bahl, and Maurizio Bozzi

Microstrip Antenna Design, K.C Gupta and Abdelaziz Benalla

Microwave and Millimeter-Wave Remote Sensing for Security Applications, Jeffrey A. Nanzer

Microwave Cavity Antennas, Akhileswar Kumar and Hristo D. Hristov

Millimeter Wave Microstrip and Printed Circuit Antennas, Prakash Bhartia

Mode-Inspired Wideband and Multiband Planar Antennas, Neng-Wu Liu, Xiao Zhang, and Lei Zhu

Mobile Antenna Systems Handbook, Third Edition, Kyohei Fujimoto, editor

Modern Automotive Antenna Measurements, Lars J. Foged and Manuel Sierra Castañer, editors

Moment Methods in Antennas and Scattering, Robert C. Hansen

Multiband Integrated Antennas for 4G Terminals, David A. Sánchez-Hernández, editor

Near-Field Antenna Measurements, Dan Slater

Noise Temperature Theory and Applications for Deep Space Communications Antenna Systems, Tom Y. Otoshi

Over-the-Air Measurement for Wireless Communication Systems, Yihong Qi and James L. Drewniak

Phased Array Antenna Handbook, Third Edition, Robert J. Mailloux

Phased Array Antennas, Arthur A. Oliner

Phased Array Antennas with Optimized Element Patterns, Sergei P. Skobelev

Plasma Antennas, Second Edition, Theodore Anderson

Polarization in Electromagnetic Systems, Second Edition, Warren Stutzman

Practical Antenna Design for Wireless Products, Henry Lau

Practical Microstrip and Printed Antenna Design, Anil Pandey

Practical Phased Array Antenna Systems, Eli Brookner

Practical Simulation of Radar Antennas and Radomes, Herbert L. Hirsch

Printed MIMO Antenna Engineering, Mohammad S. Sharawi

Radio Propagation in the Urban Scenario, Giorgio Franceschetti, Antonio Iodice, and Daniele Riccio

Radiowave Propagation and Antennas for Personal Communications, Third Edition, Kazimierz Siwiak

Reconfigurable Antenna Design and Analysis, Mohammod Ali

Reflectarray Antennas: Analysis, Design, Fabrication and Measurement, Jafar Shaker, Mohammad Reza Chaharmir, and Jonathan Ethier

Reflector Lens Antennas: Analysis and Design Using Personal Computers, Carlyle J. Sletten, editor

RF Coaxial Slot Radiators: Modeling, Measurements, and Applications, Kok Yeow You

Shipboard Antennas, Second Edition, Preston Law

Significant Phased Array Papers, Robert C. Hansen

Smart Antenna Engineering, Ahmed El Zooghby

Solid Dielectric Horn Antennas, Carlos Salema, Carlos Fernandes, and Rama Kant Jha

Sparse Phased Array Antennas: Theory and Applications, Ashutosh Kedar

Ultrawideband Antennas for Microwave Imaging Systems, Tayeb A. Denidni and Gijo Augustin

Ultrawideband Short-Pulse Radio Systems, V. I. Koshelev, Yu. I. Buyanov, and V. P. Belichenko

Waveguide Components for Antenna Feed Systems: Theory and CAD, Jaroslaw Uher, Jens Bornemann, and Uwe Rosenberg

Wearable Antennas and Electronics, Asimina Kiourti and John L. Volakis, editors

For further information on these and other Artech House titles, including previously considered out-of-print books now available through our In-Print-Forever® (IPF®) program, contact:

Artech House
685 Canton Street
Norwood, MA 02062
Phone: 781-769-9750
Fax: 781-769-6334
e-mail: artech@artechhouse.com

Artech House
16 Sussex Street
London SW1V HRW UK
Phone: +44 (0)20 7596-8750
Fax: +44 (0)20 7630 0166
e-mail: artech-uk@artechhouse.com

Find us on the World Wide Web at: www.artechhouse.com